EFFICIENCY
AND SUSTAINABILITY
IN BIOFUEL PRODUCTION

Environmental
and Land-Use Research

EFFICIENCY
AND SUSTAINABILITY
IN BIOFUEL PRODUCTION

Environmental
and Land-Use Research

Edited by
Barnabas Gikonyo, PhD

Apple Academic Press Inc. | Apple Academic Press Inc.
3333 Mistwell Crescent | 9 Spinnaker Way
Oakville, ON L6L 0A2 | Waretown, NJ 08758
Canada | USA

©2015 by Apple Academic Press, Inc.

First issued in paperback 2021

Exclusive worldwide distribution by CRC Press, a member of Taylor & Francis Group

No claim to original U.S. Government works

ISBN 13: 978-1-77463-552-0 (pbk)
ISBN 13: 978-1-77188-131-9 (hbk)

Library and Archives Canada Cataloguing in Publication

Efficiency and sustainability in biofuel production: environmental and land-use research/ edited by Barnabas Gikonyo, PhD.

Includes bibliographical references and index.
ISBN 978-1-77188-131-9 (bound)
1. Biomass energy--Environmental aspects. 2. Land use--Environmental aspects.
I. Gikonyo, Barnabas, editor

TD195.B56E34 2015   333.95'39   C2015-900212-5

Library of Congress Cataloging-in-Publication Data

Efficiency and sustainability in biofuel production: environmental and land-use research / editor, Barnabas Gikonyo, PhD.

pages cm
Includes bibliographical references and index.
ISBN 978-1-77188-131-9 (alk. paper)
1. Biomass energy. 2. Land use--Planning. I. Gikonyo, Barnabas, editor.

TP339.E25 2015   662'.88--dc23   2014049943

About the Editor

BARNABAS GIKONYO, PhD

Barnabas Gikonyo graduated from Southern Illinois University, Carbondale, Illinois (2007), with a PhD in organic and materials chemistry. He currently teaches organic and general chemistry classes at the State University of New York Geneseo, along with corresponding laboratories and the oversight of general chemistry labs. His research interests range from the application of various biocompatible, polymeric materials as "biomaterial bridging surfaces" for the repair of spinal cord injuries, to the use of osteoconductive cements for the repair of critical sized bone defects/fractures. Currently, he is studying the development of alternative, non-food biofuels.

Contents

Acknowledgment and How to Cite.. *ix*

List of Contributors.. *xi*

Introduction...*xv*

Part I: Land Use and Biofuels

1. **The Importance of Land Use Change in The Environmental Balance of Biofuels** 3

    Wassim Ben Aoun, Benoît Gabrielle, and Bruno Gagnepain

2. **Energy Potential and Greenhouse Gas Emissions from Bioenergy Cropping Systems on Marginally Productive Cropland**....................... 33

    Marty R. Schmer, Kenneth P. Vogel, Gary E. Varvel,
    Ronald F. Follett, Robert B. Mitchell, and Virginia L. Jin

3. **Integration of Farm Fossil Fuel Use with Local Scale Assessments of Biofuel Feedstock Production in Canada**... 51

    J. A. Dyer, R. L. Desjardins, B. G. McConkey, S. Kulshreshtha, and X. P. C. Vergé

4. **Evaluating the Marginal Land Resources Suitable for Developing Bioenergy in Asia** 83

    Jingying Fu, Dong Jiang, Yaohuan Huang, Dafang Zhuang, and Wei Ji

5. **Energy Potential of Biomass from Conservation Grasslands in Minnesota, USA**.. 101

    Jacob M. Jungers, Joseph E. Fargione, Craig C. Sheaffer, Donald L. Wyse, and
    Clarence Lehman

6. **Seasonal Energy Storage Using Bioenergy Production from Abandoned Croplands** 131

    J. Elliott Campbell, David B. Lobell, Robert C. Genova, Andrew Zumkehr, and
    Christopher B. Field

Part II: Second-Generation Biofuels and Sustainability

7. **Biodiesel from Grease Interceptor to Gas Tank** 151

    Alyse Mary E. Ragauskas, Yunqiao Pu, and Art J. Ragauskas

8. Efficient Extraction of Xylan from Delignified Corn Stover Using
    Dimethyl Sulfoxide 175

    John Rowley, Stephen R. Decker, William Michener,
    and Stuart Black

9. The Possibility of Future Biofuels Production Using Waste Carbon
    Dioxide and Solar Energy 187

    Krzysztof Biernat, Artur Malinowski, and Malwina Gnat

10. Well-to-Wheels Energy Use and Greenhouse Gas Emissions
    of Ethanol from Corn, Sugarcane and Cellulosic Biomass
    for US Use 249

    Michael Wang, Jeongwoo Han, Jennifer B. Dunn, Hao Cai,
    and Amgad Elgowainy

11. Lessons from First Generation Biofuels and Implications for
    the Sustainability Appraisal of Second Generation Biofuels 281

    Alison Mohr and Sujatha Raman

Author Notes 311
Index 315

Acknowledgment and How to Cite

The editor and publisher thank each of the authors who contributed to this book. The chapters in this book were previously published in various places in various formats. To cite the work contained in this book and to view the individual permissions, please refer to the citation at the beginning of each chapter. Each chapter was read individually and carefully selected by the editor; the result is a book that provides a nuanced look at efficiency and sustainability in the production of biofuels. The chapters included are broken into two sections, which describe the following topics:

- The articles in part 1 of this volume were chosen for their discussion of various topics having to do with land-use considerations, including cropping systems, marginal land resources, and seasonal energy storage from abandoned croplands.
- The articles in part 2 were selected for their novel analysis and investigations of methods and technologies for making biofuel production ever more sustainable. Included here are alternative options for biofuel feedstocks, such as recycled grease; more efficient methodologies for extracting xylan from biomass; the possibility of combining biofuel production with solar energy and waste carbon dioxide; and lessons to be learned from first-generation biofuels.

List of Contributors

Wassim Ben Aoun
AgroParisTech, INRA, UMR 1091 Environnement et Grandes Cultures, 78850 Thiverval-Grignon, France

Krzysztof Biernat
Automotive Industry Institute, Department for Fuels and Renewable Energy, Warsaw, Poland

Stuart Black
National Renewable Energy Laboratory, Golden, CO, USA

Hao Cai
Systems Assessment Group, Energy Systems Division, Argonne National Laboratory, 9700 South Cass Avenue, Argonne, IL 60439, USA

J. Elliott Campbell
School of Engineering, University of California, Merced, CA 95343, USA

Stephen R. Decker
National Renewable Energy Laboratory, Golden, CO, USA

R. L. Desjardins
Agriculture & Agri-Food Canada, Ottawa, Canada

Jennifer B. Dunn
Systems Assessment Group, Energy Systems Division, Argonne National Laboratory, 9700 South Cass Avenue, Argonne, IL 60439, USA

J. A. Dyer
Agro-environmental Consultant, Cambridge, Ontario, Canada

Amgad Elgowainy
Systems Assessment Group, Energy Systems Division, Argonne National Laboratory, 9700 South Cass Avenue, Argonne, IL 60439, USA

Joseph E. Fargione
The Nature Conservancy, Minneapolis, Minnesota, United States of America

Christopher B. Field
Department of Global Ecology, Carnegie Institutions for Science, Stanford, CA 94305, USA

Ronald F. Follett
Soil-Plant Nutrient Research Unit, United States Department of Agriculture-Agricultural Research Service (USDA-ARS), Ft. Collins, Colorado, United States of America

Jingying Fu
State Key Laboratory of Resources and Environmental Information Systems, Institute of Geographical Sciences and Natural Resources Research, Chinese Academy of Sciences, Beijing 100101, China and University of Chinese Academy of Sciences, Beijing 100049, China

Benoît Gabrielle
AgroParisTech, INRA, UMR 1091 Environnement et Grandes Cultures, 78850 Thiverval-Grignon, France

Bruno Gagnepain
Service Bioressources, Agence de l'Environnement et de la Maîtrise de l'Energie (ADEME), 20, avenue du Grésillé, BP 90406, 49004 Angers Cedex 01, France

Robert C. Genova
Department of Global Ecology, Carnegie Institutions for Science, Stanford, CA 94305, USA

Malwina Gnat
Institute for Ecology and Bioethics, University Cardinal Stefan Wyszynski, Warsaw, Poland

Jeongwoo Han
Systems Assessment Group, Energy Systems Division, Argonne National Laboratory, 9700 South Cass Avenue, Argonne, IL 60439, USA

Yaohuan Huang
State Key Laboratory of Resources and Environmental Information Systems, Institute of Geographical Sciences and Natural Resources Research, Chinese Academy of Sciences, Beijing 100101, China

Wei Ji
State Key Laboratory of Resources and Environmental Information Systems, Institute of Geographical Sciences and Natural Resources Research, Chinese Academy of Sciences, Beijing 100101, China and University of Chinese Academy of Sciences, Beijing 100049, China

Dong Jiang
State Key Laboratory of Resources and Environmental Information Systems, Institute of Geographical Sciences and Natural Resources Research, Chinese Academy of Sciences, Beijing 100101, China

Virginia L. Jin
Agroecosystem Management Research Unit, United States Department of Agriculture-Agricultural Research Service (USDA-ARS), Lincoln, Nebraska, United States of America

Jacob M. Jungers
Conservation Biology Graduate Program, University of Minnesota, Saint Paul, Minnesota, United States of America

S. Kulshreshtha
University of Saskatchewan, Saskatoon, Canada

Clarence Lehman
College of Biological Sciences, University of Minnesota, Saint Paul, Minnesota, United States of America

David B. Lobell
Environmental Earth System Science Department, Stanford University, Stanford, CA 94305, USA

Artur Malinowski
Automotive Industry Institute, Department for Fuels and Renewable Energy, Warsaw, Poland

B. G. McConkey
Agriculture & Agri-Food Canada, Swift Current, Canada

William Michener
National Renewable Energy Laboratory, Golden, CO, USA

Robert B. Mitchell
Grain, Forage and Bioenergy Research Unit, United States Department of Agriculture-Agricultural Research Service (USDA-ARS), Lincoln, Nebraska, United States of America

Alison Mohr
Institute for Science and Society (ISS), School of Sociology and Social Policy, Law and Social Sciences Building, University Park, University of Nottingham, Nottingham NG7 2RD, UK

Yunqiao Pu
Institute of Paper Science and Technology, Georgia Institute of Technology, Atlanta, Georgia

Art J. Ragauskas
School of Chemistry and Biochemistry, Georgia Institute of Technology, Atlanta, Georgia

Alyse Mary E. Ragauskas
Franklin College of Arts and Sciences, University of Georgia, Athens, Georgia

Sujatha Raman
Institute for Science and Society (ISS), School of Sociology and Social Policy, Law and Social Sciences Building, University Park, University of Nottingham, Nottingham NG7 2RD, UK

John Rowley
University of Colorado, Boulder, CO, USA

Marty R. Schmer
Agroecosystem Management Research Unit, United States Department of Agriculture-Agricultural Research Service (USDA-ARS), Lincoln, Nebraska, United States of America

Craig C. Sheaffer
Department of Agronomy and Plant Genetics, University of Minnesota, Saint Paul, Minnesota, United States of America

Gary E. Varvel
Agroecosystem Management Research Unit, United States Department of Agriculture-Agricultural Research Service (USDA-ARS), Lincoln, Nebraska, United States of America

X. P. C. Vergé
Agro-environmental Consultant to AAFC, Ottawa, Ontario, Canada

Kenneth P. Vogel
Grain, Forage and Bioenergy Research Unit, United States Department of Agriculture-Agricultural Research Service (USDA-ARS), Lincoln, Nebraska, United States of America

Michael Wang
Systems Assessment Group, Energy Systems Division, Argonne National Laboratory, 9700 South Cass Avenue, Argonne, IL 60439, USA

Donald L. Wyse
Department of Agronomy and Plant Genetics, University of Minnesota, Saint Paul, Minnesota, United States of America

Dafang Zhuang
State Key Laboratory of Resources and Environmental Information Systems, Institute of Geographical Sciences and Natural Resources Research, Chinese Academy of Sciences, Beijing 100101, China

Andrew Zumkehr
School of Engineering, University of California, Merced, CA 95343, USA

Introduction

The world's interest in reducing petroleum use has led to the rapid development of the biofuels industry over the past decade or so. However, there is increasing concern over how current food-based biofuels affect both food security and the environment. The answer is not a simple equation where biofuels automatically equal the solution to the world's energy demands in a sustainable and environmentally secure way. First-generation biofuels have the potential to be as damaging to the environment as petroleum-based products, and they may have other unintended consequences, such as reducing the availability of food in regions of the world where it is most desperately needed. That does not, however, indicate that we should throw out the baby with the bathwater, as it were. It most emphatically does not mean that the world should throw up its collective hands and say, "Well, that didn't work. Let's go back to petroleum." The answer to the problem of building a sustainable future may still rely, at least in part, on biofuels—but that will likely mean that we rely on second-generation biofuels that use widely available sources such as non-food lignocellulosic-based biomass and fats, oils, and greases. We will also need to take a closer look at how land use can simultaneously support both the world's food needs and some of its energy needs.

The purpose of this compendium volume is to consolidate research into these questions. The chapters contained here focus on three categories of research: 1) the problems currently connected with biofuels relating to land use and the environment; 2) investigations into the potential for land use to be managed effectively and sustainably; and 3) research pertaining to second-generation biofuels.

The answers to the world's environmental and energy issues are urgent. Biofuels hold great promise for all of us who are seeking to build a sustainable future.

Barnabas Gikonyo

The potential of first generation biofuels to mitigate climate change is still largely debated in the scientific and policy-making arenas. It is currently assessed through life cycle assessment (LCA), a method for accounting for the greenhouse gas (GHG) emissions of a given product from "cradle-to-grave", which is widely used to aid decision making on environmental issues. Although LCA is standardized, its application to biofuels leads to inconclusive results often fraught by a high variability and uncertainty. This is due to differences in quantifying the environmental impacts of feedstock production, and the difficulties encountered when considering land use changes (LUC) effects. The occurrence of LUC mechanisms is in part the consequence of policies supporting the use of biofuels in the trans-port sector, which implicitly increases the competition between various possible uses of land worldwide. In Chapter 1, Ben Aoun and colleagues review the methodologies recently put forward to include LUC effects in LCAs, and examples from the US, Europe and France. These cross analy-sis show that LCA needs to be adapted and combined to other tools such as economic modeling in order to provide a more reliable assessment of the biofuels chains.

Low-carbon biofuel sources are being developed and evaluated in the United States and Europe to partially offset petroleum transport fuels. In Chapter 2, Schmer and colleagues evaluated current and potential biofu-el production systems from a long-term continuous no-tillage corn (*Zea mays L.*) and switchgrass (*Panicum virgatum L.*) field trial under differ-ing harvest strategies and nitrogen (N) fertilizer intensities to determine overall environmental sustainability. Corn and switchgrass grown for bio-energy resulted in near-term net greenhouse gas (GHG) reductions of −29 to −396 grams of CO_2 equivalent emissions per megajoule of ethanol per year as a result of direct soil carbon sequestration and from the adoption of integrated biofuel conversion pathways. Management practices in switch-grass and corn resulted in large variation in petroleum offset potential. Switchgrass, using best management practices produced 3919 ± 117 liters of ethanol per hectare and had 74 ± 2.2 gigajoules of petroleum offsets per hectare which was similar to intensified corn systems (grain and 50% resi-due harvest under optimal N rates). Co-locating and integrating cellulosic biorefineries with existing dry mill corn grain ethanol facilities improved net energy yields (GJ ha^{-1}) of corn grain ethanol by >70%. A multi-feedstock, landscape approach coupled with an integrated biorefinery would be a viable option to meet growing renewable transportation fuel demands while improving the energy efficiency of first generation biofuels.

Chapter 3, by Dyer and colleagues, aims to determine the geographic distribution of farm energy terms within each province of Canada. Due to their small sizes and limited role in Canadian agriculture, the four Atlantic Provinces were treated as one combined province. A secondary goal of this chapter was to demonstrate how much the farm energy budget contributes to the GHG emissions budget of the agricultural sector through fossil CO_2 emissions at a provincial scale. Using area based intensity, a simple demonstration was also provided of how these data could provide a baseline comparison for the fossil CO_2 emitted from growing a grain ethanol feedstock compared to current types of farms. These goals were achieved through the integration of existing models and databases, rather than by analysis of new data collected specifically for this purpose.

Bioenergy from energy plants is an alternative fuel that is expected to play an increasing role in fulfilling future world energy demands. Because cultivated land resources are fairly limited, bioenergy development may rely on the exploitation of marginal land. Chapter 4, by Fu and colleagues, focused on the assessment of marginal land resources and biofuel potential in Asia. A multiple factor analysis method was used to identify marginal land for bioenergy development in Asia using multiple datasets including remote sensing-derived land cover, meteorological data, soil data, and characteristics of energy plants and Geographic Information System (GIS) techniques. A combined planting zonation strategy was proposed, which targeted three species of energy plants, including *Pistacia chinensis* (*P. chinensis*), *Jatropha curcas* L. (*JCL*), and *Cassava*. The marginal land with potential for planting these types of energy plants was identified for each $1 \, km^2$ pixel across Asia. The results indicated that the areas with marginal land suitable for *Cassava, P. chinensis*, and *JCL* were established to be 1.12 million, 2.41 million, and 0.237 million km^2, respectively. Shrub land, sparse forest, and grassland are the major classifications of exploitable land. The spatial distribution of the analysis and suggestions for regional planning of bioenergy are also discussed.

Perennial biomass from grasslands managed for conservation of soil and biodiversity can be harvested for bioenergy. Until now, the quantity and quality of harvestable biomass from conservation grasslands in Minnesota, USA, was not known, and the factors that affect bioenergy potential from these systems have not been identified. In Chapter 5, Jungers and colleagues measured biomass yield, theoretical ethanol conversion efficiency, and plant tissue nitrogen (N) as metrics of bioenergy potential from mixed-species conservation grasslands harvested with commercial-scale equipment. With three years of data, the authors used mixed-effects models to determine factors that influence bioenergy potential. Sixty conservation grassland plots, each about 8 ha in size, were distributed among three locations in Minnesota. Harvest treatments were applied annually in autumn as a completely randomized block design. Biomass yield ranged from 0.5 to 5.7 Mg ha^{-1}. May precipitation increased biomass yield while precipitation in all other growing season months showed no affect. Averaged across all locations and years, theoretical ethanol conversion efficiency was 450 l Mg^{-1} and the concentration of plant N was 7.1 g kg^{-1}, both similar to dedicated herbaceous bioenergy crops such as switchgrass. Biomass yield did not decline in the second or third year of harvest. Across years, biomass yields fluctuated 23% around the average. Surprisingly, forb cover was a better predictor of biomass yield than warm-season grass with a positive correlation with biomass yield in the south and a negative correlation at other locations. Variation in land ethanol yield was almost exclusively due to variation in biomass yield rather than biomass quality; therefore, efforts to increase biomass yield might be more economical than altering biomass composition when managing conservation grasslands for ethanol production. These measurements of bioenergy potential, and the factors that control it, can serve as parameters for assessing the economic viability of harvesting conservation grasslands for bioenergy.

Bioenergy has the unique potential to provide a dispatchable and carbon-negative component to renewable energy portfolios. However, the sustainability, spatial distribution, and capacity for bioenergy are critically dependent on highly uncertain land-use impacts of biomass agriculture. Biomass cultivation on abandoned agriculture lands is thought to reduce land-use impacts relative to biomass production on currently used croplands. While coarse global estimates of abandoned agriculture lands have been used for large-scale bioenergy assessments, more practical technological and policy applications will require regional, high-resolution information on land availability. In Chapter 6, Campbell and colleagues present US county-level estimates of the magnitude and distribution of abandoned cropland and potential bioenergy production on this land using remote sensing data, agriculture inventories, and land-use modeling. These abandoned land estimates are 61% larger than previous estimates for the US, mainly due to the coarse resolution of data applied in previous studies. The authors apply the land availability results to consider the capacity of biomass electricity to meet the seasonal energy storage requirement in a national energy system that is dominated by wind and solar electricity production. Bioenergy from abandoned croplands can supply most of the seasonal storage needs for a range of energy production scenarios, regions, and biomass yield estimates. These data provide the basis for further down-scaling using models of spatially gridded land-use areas as well as a range of applications for the exploration of bioenergy sustainability.

The need for sustainable biofuels has initiated a global search for innovative technologies that can sustainably convert nonfood bioresources to liquid transportation fuels. While 2nd generation cellulosic ethanol has begun to address this challenge, other resources including yellow and brown grease are rapidly evolving commercial opportunities that are addressing regional biodiesel needs. Chapter 7, by Ragauskas and colleagues, examines the technical and environmental factors driving the collection of trap FOG (Fats, Oils, and Greases), its chemical composition and technologies currently available and future developments that facilitate the conversion of FOG into biodiesel.

Xylan can be extracted from biomass using either alkali (KOH or NaOH) or dimethyl sulfoxide (DMSO); however, DMSO extraction is the only method that produces a water-soluble xylan. In Chapter 8, Rowley and colleagues studied the DMSO extraction of corn stover at different temperatures with the objective of finding a faster, more efficient extraction method. The temperature and time of extraction were compared followed by a basic structural analysis to ensure that no significant structural changes occurred under different temperatures. The resulting data showed that heating to 70 °C during extraction can give a yield comparable to room temperature extraction while reducing the extraction time by ~90 %. This method of heating was shown to be the most efficient method currently available and was shown to retain the important structural characteristics of xylan extracted with DMSO at room temperature.

Chapter 9, by Biernat and colleagues, argues that the most important goals for the energy sector should be utilization of carbon dioxide emissions. This target can be realized by using modern technology of CO_2 reuse to produce liquid energy carriers.

Globally, bioethanol is the largest volume biofuel used in the transportation sector, with corn-based ethanol production occurring mostly in the US and sugarcane-based ethanol production occurring mostly in Brazil. Advances in technology and the resulting improved productivity in corn and sugarcane farming and ethanol conversion, together with biofuel policies, have contributed to the significant expansion of ethanol production in the past 20 years. These improvements have increased the energy and greenhouse gas (GHG) benefits of using bioethanol as opposed to using petroleum gasoline. Chapter 10, by Wang and colleagues, presents results from our most recently updated simulations of energy use and GHG emissions that result from using bioethanol made from several feedstocks. The results were generated with the GREET (Greenhouse gases, Regulated Emissions, and Energy use in Transportation) model. In particular, based on a consistent and systematic model platform, we estimate life-cycle energy consumption and GHG emissions from using ethanol produced from five feedstocks: corn, sugarcane, corn stover, switchgrass and miscanthus The authors quantitatively address the impacts of a few critical factors that affect life-cycle GHG emissions from bioethanol. Even when the highly debated land use change GHG emissions are included, changing from corn to sugarcane and then to cellulosic biomass helps to significantly increase the reductions in energy use and GHG emissions from using bioethanol. Relative to petroleum gasoline, ethanol from corn, sugarcane, corn stover, switchgrass and miscanthus can reduce life-cycle GHG emissions by 19–48%, 40–62%, 90–103%, 77–97% and 101–115%, respectively. Similar trends have been found with regard to fossil energy benefits for the five bioethanol pathways.

The emergence of second generation (2G) biofuels is widely seen as a sustainable response to the increasing controversy surrounding the first generation (1G). Yet, sustainability credentials of 2G biofuels are also being questioned. Drawing on work in Science and Technology Studies, in

Chapter 11, Mohr and Raman argue that controversies help focus attention on key, often value-related questions that need to be posed to address
broader societal concerns. This paper examines lessons drawn from the
1G controversy to assess implications for the sustainability appraisal of
2G biofuels. The authors present an overview of key 1G sustainability
challenges, assess their relevance for 2G, and highlight the challenges
for policy in managing the transition. The article addresses limitations of
existing sustainability assessments by exploring where challenges might
emerge across the whole system of bioenergy and the wider context of the
social system in which bioenergy research and policy are done. Key lessons arising from 1G are potentially relevant to the sustainability appraisal
of 2G biofuels depending on the particular circumstances or conditions
under which 2G is introduced. The article concludes that sustainability
challenges commonly categorised as either economic, environmental or
social are, in reality, more complexly interconnected (so that an artificial
separation of these categories is problematic).

PART I

LAND USE AND BIOFUELS

CHAPTER 1

The Importance of Land Use Change in the Environmental Balance of Biofuels

WASSIM BEN AOUN, BENOOT GABRIELLE, AND BRUNO GAGNEPAIN

1.1 INTRODUCTION

The use of bioenergy in the transport sector is one of the solutions proposed by policy-makers to mitigate climate change and promote energy security. In the short to medium term, the European Union (EU) aims to deploy first generation biofuels, especially biodiesel and ethanol in order to replace fossil fuel and reduce anthropogenic emissions of greenhouse gases (GHG). Earlier studies (Farrel et al., 2006; Wang, 2005) concluded to a significant abatement of GHG emissions when substituting petroleum-based fuels with biofuels, which prompted the development of biodiesel and ethanol. However, recent pieces of research (Fargione et al., 2008; Searchinger et al., 2008) have suggested that policies supporting

The Importance of Land Use Change in the Environmental Balance of Biofuels © Ben Aoun W, Gabrielle B, and Gagnepain B. Oilseeds and Fats, Crops and Lipids, 20,5 (2013), http://dx.doi.org/10.1051/ocl/2013027.

biofuels deployment should be revised in order to limit the unintended impacts of biofuel expansion, whereby the displacement of food crops by energy crops not only leads to direct land use changes (dLUC) but also to indirect land uses changes (iLUC). These complex mechanisms are difficult to estimate and are usually associated with detrimental effects on the environment, such as increased emissions of GHG and biodiversity depletion from the conversion of natural ecosystems. Thus, they are likely to severely degrade the environmental performance of biofuels.

Life-cycle assessment (LCA) is currently the most widely-used method to assess the environmental sustainability of biofuels, in particular for policy-making purposes. However, most published LCA studies on biofuels do not take into account iLUC effects (Di Lucia et al., 2012). This is in fact due to the inability of classical (also called attributional) LCA to take into account such effects, since it ignores the market and economic implications of a given decision (eg, to achieve a given blending target for biofuels at a national level). Economic equilibrium models and the so-called consequential approach to LCA have been promoted as a more suitable alternative to include these effects and produce a robust assessment of biofuels environmental impacts (Kløverpris et al., 2008). Although there is a consensus on the fact that LUC effects need to be addressed, the resulting indicators are quite heterogeneous and subject to high uncertainty (De Cara et al., 2012).

The EU is increasingly concerned with this issue and is currently expecting more reliable results to frame its biofuel policy. A directive on renewable energies was released by the European Commission (EC) in 2009, introducing sustainability criteria to be assessed when producing biofuels (EC, 2009; EC, 2010). However, its support of first-generation biofuels was recently questioned (EC, 2010).

The goal of this study is to underline the importance of using LCAs to evaluate the environmental burdens associated to biofuels chains, and the necessity of adapting this methodology in order to allow taking into account the effects of LUC. This work also focuses on identifying the sources of variability and uncertainty of existing studies.

FIGURE 1: Fossil fuel vs. biofuel life cycles (Wang, 2005).

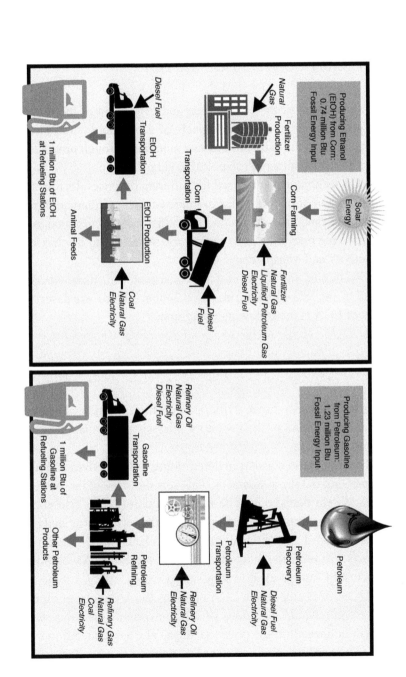

1.2 LCA: A SUITABLE TOOL FOR ENVIRONMENTAL ASSESSMENT

1.2.1 MAIN CONCEPTS OF AN LCA

LCA is defined as the compilation and evaluation of the inputs, outputs and potential environmental impacts of a product system throughout its life cycle from the extraction of raw materials through production and use to waste management (Curran, 2013) (Fig. 1).

LCA technique can be used for different purposes. Its results allow the identification of opportunities to improve the environmental performance of products and provide a sound scientific basis for decision makers. This is due to the relevance of its indicators, as well as to its characteristics of objectivity and transparency.

Conducting the LCA must be consistent with the methodology proposed by the ISO 14040 series (ISO, 2006a; 2006b). We describe the main steps of LCA in the following paragraph.

1.2.2 STEPS OF AN LCA

LCA is an iterative process divided into four interrelated stages: the goal and scope definition, the inventory analysis, the impact assessment, and the interpretation.

The first phase consists in determining the objectives and the rationale for carrying out the assessment. This sets the scale of the study and establishes system boundaries. The functional unit (FU) is also chosen during this step. It measures the performance of the service provided by the product and is used as reference unit when calculating all the environmental impacts. For biofuels, functional units are typically 1 MJ of biofuel energy content or 1 km travelled in a passenger car.

The second step in the LCA involves the inventory of inputs, outputs and environmental emissions of all components (or subsystems) of the system delineated in the previous stage. The associated flows (of materials, energy, information, etc.) are listed for each subsystem and expressed on the basis of the FU.

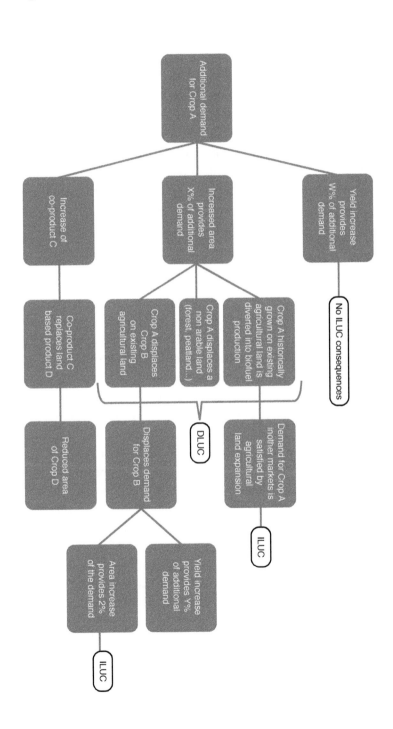

FIGURE 2: Consequences of an increase in biofuels production (adapted from Bauen et al., 2010).

The impact analysis phase assesses the environmental impacts of inputs and outputs of the system studied, by aggregating the flows of the life cycle inventory into a set of impact categories, weighing all substances relative to a reference substance for each of these categories. For instance, the reference substance for the global warming impact is carbon dioxide (CO_2), and nitrous oxide (another greenhouse gas) will be given a weight of 296 corresponding to its global warming potential relative to CO_2 (ADEME, 2010). Interpretation phase is a key stage in which the robustness of the results is evaluated. This allows determining the main conclusions, limitations and recommendations borne out of the LCA study.

1.2.3 APPLICATION TO BIOFUELS: ATTRIBUTIONAL VS. CONSEQUENTIAL LCA

In the recent literature on LCA, two approaches are distinguished. Attributional, also called retrospective LCA (aLCA) provides information about the environmental properties of a particular life cycle, and its subsystems. It thus seeks to describe the environmental impacts of past, current or potential future product systems, independent of other products or systems that could be affected by their development. Consequential, also called prospective LCA (cLCA) provides information on the environmental consequence of individual actions, eg the deployment of such products (Ekvall and Weidema, 2004). In aLCA, the system investigated is restricted to a single life cycle from cradle to grave. Technical data on the various sub-systems of the life-cycle are averaged across the geographical domain considered to determine mean environmental burdens per unit of product considered.

Co-products associated with the product of interest are handled by applying allocation factors or using system expansion (Wang et al., 2011). In the case of biofuels, it is thus permitted to allocate a portion of the environmental burdens due to agricultural feedstock production and the first steps of the industrial processing to the co-products. The energy allocation remains the most commonly used method in the handling of co-products (Wang et al., 2011).

At this level, aLCA seems able to provide a comprehensive assessment of biofuels life cycles, since all the effects directly resulting from their production are taken into account (Reinhard and Zah, 2011). However, with the development of first generation biofuels sector in Europe and in the United States (US), it was found that their production entails large-scale modifications of terrestrial ecosystems and biospheric fluxes through indirect market mechanisms (Fargione et al., 2008; Mellilo et al., 2009; Searchinger et al., 2009).

Consequential LCA can address these effects by simulating a "shock" in biofuel demand. It expands the system to include the life cycles of products affected by a change of the physical flows in the central life cycle. So it analyzes the system beyond the classical boundaries of the biodiesel value chain (from feedstock production to combustion in a vehicle), by encompassing the effects of fossil fuel substitution on other sectors or markets (e.g. food and feed commodities). In addition, regarding co-products, cLCA avoids allocation and thus should ideally model displacement of alternative products as a result of dynamic market interactions. Consequential LCA relies on marginal data as opposed to average data for aLCA (Ekvall et al., 2005).

The study of Searchinger, et al. (2008) illustrates the variation of LCA results in the case of biofuels according to the approach used. The results obtained with the attributional approach encourage the development of bioethanol from corn-based ethanol in the US, while those from the consequential approach point to an increase in GHG emissions if ethanol is to substitute gasoline. This difference is mainly due to the inclusion of the iLUC effects and the conversion of natural ecosystems to arable land.

1.3 LAND USE CHANGES DUE TO BIOFUELS DEVELOPMENT

Taking both the direct and indirect effects of biofuels development into account is essential to improve the environmental assessment. Is the use of LCA methodology sufficient to provide an accurate estimation of the biofuels environmental performance?

Changes in land-use are the most important consequence of biofuels production (Van Stappen et al., 2011). Thus, the environmental assessment quality depends strongly on the way in which the magnitude of these mechanisms and their environmental effects are measured.

In this section we review current knowledge of LUC in relation to biofuel development, with a focus on their complexity and on the characteristics of each type of LUC. We also present the different methods used for their estimation.

1.3.1 TYPES OF LAND USE CHANGES

The development of biofuels creates additional opportunities for economic agents. In fact, the increase in demand for a given biofuel feedstock will create a shortage of this product, which increases its price and thus provides an incentive for the farmer to increase its production. Farmers respond to this situation by intensifying crop management to improve yields. They may also transform uncultivated lands (e.g. natural areas, fallow) into arable land and/or substitute food/feed crops by energy crops (Reinhard and Zah, 2011).

The expansion of land devoted to energy crops and the displacement of food crops trigger LUC mechanisms. Here, we distinguish two types of LUC: dLUC and iLUC.

1.3.1.1 DIRECT LAND USE CHANGES (DLUC)

Direct land use change takes place when biofuels feedstock cultivation modifies the land use (De Cara et al., 2012). According to Gawel and Ludwig (2011), this type of LUC occurs when biomass cultivation displaces a different former land use (e.g., an arable crop grown on a former grassland). For Van Stappen, et al. (2011), dLUC describes the introduction of a new cropping system in a site where this form of cultivation has not taken place before. It may be estimated quantitatively from the changes in soil and vegetation carbon stocks.

If the biofuel market only changes the valorization of a given crop (i.e. switching from a food to energy en-use), the local impacts are considered negligible. On the other hand, if energy crops displace other crops in a cropping system, the effects on environment may be significant. Lastly, if feedstock production occurs on land with high carbon stocks (e.g. pasture, peat land, unmanaged forests), the dLUC effect is expected to be adverse. Conversely, when biofuel feedstock are grown on degraded soil, dLUC can contribute to improving the soil carbon balance (Gnansounou et al., 2008).

Several studies (EC, 2009; Hamelinck et al., 2008) focused on GHG emissions due to dLUC. Their results show that GHG emissions due to dLUC can be positive or negative depending on the type of land use prior to the implementation of energy crops (Van Stappen et al., 2011).

1.3.1.2 INDIRECT LAND USE CHANGE (ILUC)

The development of first generation biofuels inevitably increases the pressure on land uses worldwide, and ultimately brings into cultivation lands that otherwise would not have been put to this use (Delucchi, 2011). iLUC occurs when additional demand for bioenergy feedstock induces a change in land use on other places via market mechanisms in order to maintain the same production level of food/feed crops (De Cara et al., 2012; Van Stappen et al., 2011). According to Gawel and Ludwig (2011), iLUC occurs when land that was formerly used for the cultivation of food, feed or fibre is now used for biomass production, shifting the original land use to an alternative area that may have a high carbon stock.

Contrary to dLUC, it is often impossible to quantify iLUC associated to bioenergy development, since it is a mechanism that can occur outside the country having fostered the production of biofuels. For example, Laurance (2007) showed that the increase of corn planting in the US may affect result in deforestation in the Amazon region. Thus, iLUC can cause important GHG emissions, with also adverse effects on biodiversity as well as on soil and water quality.

1.3.2 COMPLEXITY OF THE MECHANISMS

Theoretically, both direct and indirect LUC mechanisms appear quite simple. As shown in the previous section, any increase in biofuel production ultimately requires diverting cropland to the production of biofuel feedstock, which inevitably causes dLUC and iLUC.

For example (Fig. 2), an additional demand for rapeseed (crop A) used to produce biodiesel is met through two main market responses: an increase in current yields and an expansion in the cultivation area of rapeseed to ensure biodiesel production. With the second option (land expansion), rapeseed historically grown on existing agricultural land can be diverted to biodiesel production. This type of dLUC reduces agricultural area of rapeseed used in food, which has to be produced in some other land, incurring an iLUC effect. Rapeseed may also be grown on non-agricultural land (e.g., fallow and grassland). This type of dLUC is generally not accompanied by iLUC. Expansion of rape production can also be met by displacing other crops (crop B) on existing agricultural land. This can trigger iLUC in order to satisfy the demand in the displaced crops. One should mention that the production of biodiesel from rapeseed also allows meal production (co-product C) that can substitute other products used in animal feeding from another crop (crop D). This substitution reduces surface on which crop D is cultivated and therefore mitigates iLUC (Bauen et al., 2010).

We emphasize the importance of addressing LUC issues especially for iLUC on a global scale to allow taking into account the overall consequences of biofuels production (Di Lucia et al., 2012; Reinhard and Zah, 2009; Van Stappen et al., 2011).

The increase of pressure on land and the crops displacement that occur in major exporting countries such as Europe and the US change the market balance of products from these crops and thus affect their prices (De Cara, et al. 2012). This has an effect on farmers' decisions regarding the allocation of land worldwide.

In other words, as long as crops are displaced, the effects of displacement trickle through the overall global agriculture system until it reaches a new equilibrium (Delucchi, 2011; Reinhard and Zah, 2009).

Moreover, these new equilibria may promote substitution among several products (e.g., palm oil may substitute rapeseed oil used for biodiesel production). This makes the LUC mechanisms increasingly complex and their monitoring difficult to the point that the estimation of their environmental impacts is impossible (Overmars et al., 2011). Furthermore, one should mention that LUCs are also driven by several other factors such as biophysical, demographic and economic forces. Thus, attempting to attribute LUC to a single factor or the isolation of LUC only due to biofuels production reveals serious problems (De Cara et al., 2012; Gnansounou et al., 2009).

At European level, GHG emissions from dLUC may be assessed on the basis of the guidelines developed by the Intergovernmental Panel on Climate Change (IPCC), which propose default emissions factors (Tier 1) but also recommend using country-specific validated data (Tier 2 or 3) wherever available (Van Stappen et al., 2011). Unfortunately, the complexity of above-mentioned mechanisms leads to the fact that there is currently no consensus on one method for estimating GHG from iLUC (Gawel and Ludwig, 2011; Plevin et al., 2010), despite the general awareness that neglecting or over/under estimate of iLUC effects leads to wrong decisions and to an inefficient use of biofuels.

1.3.3 LAND USE CHANGE ESTIMATION

With the awareness of their importance in the environmental balances of biofuels, LUC effects are currently widely investigated, and estimated using different approaches, which are reviewed in the following section.

1.3.3.1 MONITORING: USE OF HISTORICAL DATA AND STATISTICAL ANALYSIS

Historical data from different sources may be collected and analyzed from a statistical viewpoint to identify possible relationships between biofuel production rates in a given country and land use and land use change. The use of this method is often justified by the fact that if biofuel production in a given country did trigger land conversion elsewhere, evidence for LUC effects should be traceable in past data on land-use worldwide (Kim and Dale, 2011; Overmars et al., 2011). Some studies have attempted to find evidence for LUC from historical data. A recent study was conducted by In Numeri (2012) on behalf of the French Agency for Environment and Energy Management (ADEME) to identify the impacts of biofuel production in France on the French and international markets (imports, exports, prices, etc.), as well as on LUC. It concludes that LUCs in France are relatively limited, but it leads to inconclusive results concerning LUCs in countries outside the European Union. Also on behalf of ADEME, Chakir and Vermont (2013) analyzed the evolution of land use and dLUC generated by the development of energy crops and food crops in France during the last two decades, based on data from annual land-use surveys TERUTI (AGRESTE, 2004) and TERUTI LUCAS (AGRESTE, 2010). This study showed that until 2004, the increase of energy and food crops areas was limited to agricultural land while from 2006 on the expansion of these surfaces also impacted permanent grassland.

This approach was used by Kim and Dale (2011) to detect evidence for iLUC that might be caused by biofuel production in the US through a statistical analysis. This kind of analysis seems not to be sensitive enough to detect iLUC due to biofuel development. In contrast, Overmars, et al. (2011) used the same approach with a set of assumptions and concluded that emissions from iLUC could shift the GHG balance for biofuels from a net abatement to a net surplus of emissions relative to fossil fuels.

These retrospective and ex-post analyses are useful to illustrate the complexity of LUC mechanisms, but usually do not allow the isolation of LUC due to biofuels development from simple statistical analyses (De Cara, el al. 2012; Di Lucia et al., 2012; Overmars et al., 2011). Assumptions (e.g. on where the iLUC is likely to occur) must be made in order to obtain some uncertain conclusions.

1.3.3.2 EXPERT BASED OPINIONS

As indicated above, statistical analyses of historical data are not sufficient to isolate and quantify the impact of biofuels production on LUC. The understanding of the mechanisms and the consultation of expert opinion remain essential to be able to locate the LUC (notably iLUC) and predict their magnitude. This method, also called "causal descriptive" is known for the transparency of its assumptions, often based on intuitive cause -effect relations, and its simplification of market mechanisms (Bauen et al., 2010; Fritsches et al., 2010; Nassar et al., 2011).

Such approaches are often used in consequential LCA. For example, Reinhard and Zah (2011) made some assumptions based on expert opinions to define a priori the crops displaced by biofuels feedstock, as well as the origin of the products to be imported to offset the decline in rapeseed oil diverted into biodiesel production in Switzerland.

1.3.3.3 ECONOMIC EQUILIBRIUM MODELS

In practice, it is impossible to isolate the impact of biofuels development on land use change from historical data or experts based opinions as there are other activities that can lead to exchanges in land use. Moreover, these methods simplify market mechanisms so that the prediction of LUC (especially of iLUC) might be not accurate enough. Actually, modeling seems to be the most successful method for measuring both direct and indirect LUC (Edwards, et al. ?). Quantitative assessments based on models have been the policy makers' preferred methodology, even if they always blame their lack of transparency compared to LCA. Today general consensus exists about using economic approach to address iLUC (Di Lucia et al., 2012). This approach consists of using economic equilibrium models, which are complex optimization models based on the assumptions of perfect markets reaching equilibrium when demand equals supply in the studied economy. The response of supply and demand to price changes is the basis of the estimate of the LUC. These models make it possible to pinpoint the consequences of an additional demand for biofuels on land use at global scale, provided they include a land-use module and some degree of spatial differentiation between world regions. Here, we separate between two types of equilibrium models: partial and general equilibrium models.

1.3.3.3.1 Partial Equilibrium Models

The partial equilibrium models address a particular economic sector. Those who represent the agriculture describe the different compartments of commodities supply (yields, areas allocated to different cultures, imports) and demand (human/animal demand, non-food demand, and exports). They estimate land demand for individual crops and allow them to compete for land through cross-price elasticity. They subsequently calculate prices that balance supply and demand in all markets represented and their evolution over time within a given time horizon (Nassar et al., 2011). The main partial equilibrium models are FAPRI, FASOM, CAPRI, IMPACT, GLOBIOM, AGLINK-COSIMO, and MIST.

1.3.3.3.2 General Equilibrium Models

General equilibrium models address all economic sectors and are developed to describe the international trade. Interactions between different markets are recognized endogenously in the model. It can be assumed that land is relatively easily transformed from one use to another through the definition of a constant elasticity of transformation. These models also include a representation of the yields response to price and make the differentiation between the yields from new lands and those from land already cultivated. Thus, the farmer will choose between increased business through the adjustment of production factors use levels (labor, fertilization, etc.), or expansion on other land to meet the demand due to the development of biofuels. The main general equilibrium models are GTAP, LEITAP, and MIRAGE (De Cara et al., 2012).

Here, one should emphasize the necessity to incorporate geo-referenced information as inputs in economic models, especially regarding land cover and land availability. Certainly, with a finer spatial resolution, the estimation of GHG from LUC is the more accurate. It is also crucial to use biophysical models in combination with economic models in order to provide necessary information on yields and GHG emissions.

1.4 AVAILABLE ESTIMATIONS FOR LUC EFFECTS

1.4.1 MAIN RESULTS

1.4.1.1 AT FRENCH LEVEL (STUDIES COMMISSIONED BY ADEME)

By means of a sensitivity analysis through a wide range of scenarios, the LCA of biofuels consumed in France (ADEME, 2010) highlighted the large sensitivity of their GHG balances to LUC hypothesis. Figure 3 shows the variation range of GHG emissions associated with different LUC hypothesis in a sensitivity analysis on GHG balance of soy biodiesel. Different bars represent gasoil GHG emissions, GHG emissions for the soy biodiesel pathway considered and the GHG emissions for the different examined LUC scenario.

To contribute to improve knowledge on this topic, ADEME decided to work in partnership with the French Rational Institute for Agricultural Research (INRA) to provide additional analysis on ways of accounting for LUC in the GHG balance evaluation. In this section, the emphasis was placed on the different studies resulting from this collaboration and shared with representatives of public bodies, technical and scientific experts, and NGOs. The first step of this partnership was the launching at the end of 2010 with an international literature review carried out by INRA and a retrospective analysis of the impacts of French biofuel development policy since the 1990's. This dual approach enabled to study this question from different perspectives:

- in a prospective way, on a variable geographic scale, with various hypothesis especially on the LUC type, on feedstock mobilized, type of biofuels, by means of an international literature review;
- in retrospect, focusing at French level to examine the impacts of a national biofuel policy on a given period, with definite biofuel pathways and LUC types.

The retrospective analysis is described below in terms of aims, scope, methodology and main results and outcomes. General trends emerging from the review of international literature are given in Section 1.4.1.2.3, while a particular focus is given to a set of key studies deemed particularly representative of current literature in Sections 1.4.1.2.1 and 1.4.1.2.2. In those studies only the evaluations of global LUC factors or GHG balances pertaining to biodiesel pathways are presented here, in line with this special issue.

Based on the above-mentioned results from the sensitivity analysis on LUC scenarios in France (ADEME, 2010), it seemed interesting to investigate whether the development of biofuel consumption in France between 1993 and 2009 could have induce impacts on French and global markets of agricultural raw materials, processed products and co-products and LUC (direct or indirect). This survey was carried out by combining complementary approaches, presented in[1]:

1. data collection, statistical analysis in order to identify correlation between data series, evaluation of areas needed for production of raw materials (In Numeri, 2012);
2. analysis of land cover and land use changes in France (Teruti and Teruti-Lucas), evaluation of GHG emissions associated to biofuel consumption development: assessment of direct LUC in France (Chakir and Vermont, 2013);
3. economic modeling at France, European and global levels with a partial equilibrium model focused on crops: investigation of LUC and iLUC.

The first study (In Numeri, 2012) mainly evidenced the growing part of imports of raw materials (oils or oilseeds) used for biodiesel production between 2006 and 2009. The resulting LUC in France appeared relatively limited, essentially corresponding to the reversal of land set-aside in 1992. In other areas of the world, contrasting situations were observed. However, the statistical analyses did not make it possible to conclude about the associated impacts in terms of GHG emissions and thus to estimate global LUC factors. This confirms the difficulty if not impossibility in the absence of modeling to determine the sole responsibility of biofuels in the evolution of cropland evolution, crop management and land use changes.

The second study (Chakir and Vermont, 2013) confirmed that the increase in cropland area dedicated to energy use (rapeseed, sunflower, wheat and sugar beet) in France between 1992 and 2010 remained limited to existing agricultural land through the cultivation of land that had been set-aside from 1992 on, and to a lesser extent the conversion of grassland to arable land. For winter rapeseed, the increased crops area was obtained through re-allocations within existing arable land. The growth in sunflower area was done at the expense of mixed areas between livestock and crops with a slightly higher conversion rate of grassland towards cropland. An attempt at evaluating a dLUC factor was made for the 2007−2010 time slice on the hypothesis that the dLUC structure was similar between the whole area cropped to rapeseed or sunflower and the area dedicated to energy feedstock. This lead to the following ranges: 0.2 to 0.6 g eq. CO_2/MJ for rapeseed biodiesel and 0.7 to 1.9 g eq. CO_2/MJ for sunflower biodiesel.

1.4.1.2 AT EUROPEAN AND GLOBAL LEVEL

1.4.1.2.1 Economic Studies

Carried out on behalf of the European Commission and published in autumn 2011, the IFPRI study (Laborde, 2011) used the economic general equilibrium model Mirage-Biof. It aimed at assessing the impacts (expressed as a dLUC + iLUC factor) of the forecast biofuel consumption patterns of the 27 member States of the EU in 2020 based on their respective National Renewable Energy Action Plans. Only first generation biofuels were considered. Table 1 lists the dLUC + iLUC factors obtained for different biodiesel pathways in two situations (without and with trade liberalization).

Several LUC studies were published in 2010 and 2011 by the Joint Research Centre of the EU (JRC). The work of Marelli, et al. (2011) is based on the evaluations of IFPRI with the same feedstock, production areas, biofuel types, biofuel demand patterns, feedstock, and time horizon. The main differences with the IFPRI study lie in the classification of certain crops as annual or perennial plants, the use of updated emission factors for some kind of soils (e.g. peatlands) and a finer categorization of available lands for cropland growth. The work of Edwards, et al. (2010) compared different economic models (FAPRI, GTAP, LEITAP), and considered different time horizons and biofuel consumption levels (Tab. 2).

US Environmental Protection Agency (EPA) also published several studies in 2009 (USEPA, 2009) and 2010 (USEPA, 2010), focused on the impacts of US biofuel consumption targets at different time scales (2012, 2017 and 2022 for ethanol and only 2022 for soy biodiesel), based on the FASOM and FAPRI models (Tab. 3).

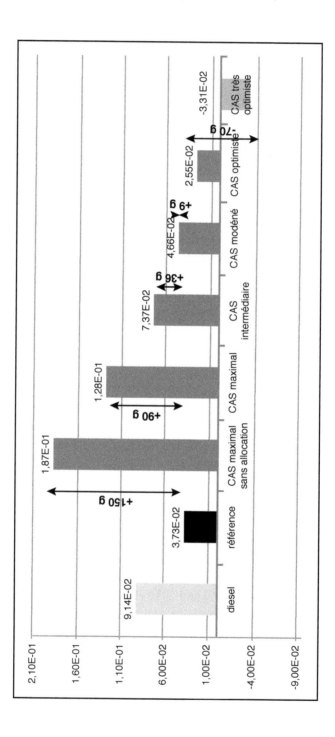

FIGURE 3: Example of LUC sensitivity analysis on GHG balance of soy biodiesel (ADEME, 2010).

TABLE 1: Direct and indirect LUC factors (g eq. CO_2/MJ), computed over a 20 year period for various biodiesel pathways (source: Laborde 2011).

Biodiesel	Without trade liberalization	With trade liberalization
Rapeseed	54	55
Sunflower	52	53
Soy	56	57
Palm	54	55

TABLE 2: Direct and indirect LUC factor in g eq. CO_2/MJ computed over a 20 year period for several biodiesel pathways.

	(Marelli et al. 2011)*	(Edwards et al. 2010)		
Biodiesel		FAPRI	GTAP	LEITAP
Rapeseed	51.6–56.6	73–221	57–73.6	338–353
Sunflower	56.2–60.4	–	–	–
Soy	51.5–55.7	–	–	–
Palm	54–55	–	14–78	75–368

Range of values corresponding to different values of soil organic carbon content.

TABLE 3: Direct and indirect LUC factor in g eq. CO_2/MJ computed over a 20 year period for soy biodiesel.

Biodiesel	USEPA 2009	USEPA 2010
Soy	154	48.5

1.4.1.2.2 LCA Studies

Compared to economic studies on LUC effects, there are far fewer references available in the international literature review based on the LCA approach. We present below only those studies which present disaggregated dLUC and iLUC factors for different biodiesel pathways. Acquaye, et al. (2011) examined the case of rapeseed-based biodiesel when meeting the 2020 target of 10% of renewable energy in transportation sector in the EU.

According to LUC type (grassland to cropland or forest to cropland), the respective estimated direct and indirect LUC factor is 26 g eq. CO_2/MJ or 53.7 g eq. CO_2/MJ.

Table 4 compiles results of several studies, dealing with the biofuel policy of a particular country in Europe, considering different LUC types for different biodiesel pathways (rapeseed, soy, and palm).

TABLE 4: Direct and indirect LUC factor in g eq. CO_2/MJ annualized on 20 years for different pathways and different national policy schemes in Europe.

d+i LUC (g eq. CO_2/ MJ)	(Lechon et al., 2011)*	(Reinhard and Zah)**	(Brandao, 2011)***	(ADEME, 2010)****
Mix biodiesel Europe	122–127	–	–	–
Rapeseed	–	–145–307	–280–380	–48–99
		(Reinhard and Zah, 2011)		
Soy	–	–85–125	–	–38–444
		(Reinhard and Zah, 2009)		
Palm	–	193	–	–11,6–120
		(Reinhard and Zah, 2009)		

*Spain, raw material supply areas: Europe, US, Canada, Malaysia – LUC types considered (grassland-cropland, forest-cropland, others), different coproduct effect levels, **Switzerland, raw material supply areas: Switzerland, Brazil, Malaysia, ***UK, raw material supply areas: UK-alternatively expansion, substitution of lands, intensification of crops, different LUC types, different biofuel consumption levels, ****France, raw material supply areas: Europe, Brazil, US, Malaysia, Indonesia, different LUC types.

1.4.1.2.3 International Literature Review

A recent study by De Cara, et al. (2012) surveyed the international literature on LUC and iLUC effects related to biofuel development, and aimed at evaluating their level and analyze their impacts on the GHG balances of biofuels. It focused on biodiesel (methyl esters) and bioethanol pathways. 485 references published between 1996 and 2011 were identified,

70 which were retained after an accurate selection, providing 239 direct LUC factors and 561 direct and indirect LUC factors.

The first conclusion drawn from this work was that LUC issue remains a recent scientific concern, which was still unknown when the French biofuel plans were launched and logically not taken account at the time.

The analysis of overall direct and indirect LUC factor shows some pretty clear differences according to raw material, biofuel types (1st vs. 2nd generation), supply area of raw materials, biofuel demand area, and methodology.

Among the 561 evaluations of overall LUC factor cited above, 221 involved biodiesel pathways, mostly based on rapeseed, soybean and palm oil. In order to get a better idea of these evaluations and potential impacts, these figures were added to the attributional life-cycle emissions of GHG of biofuels in France (ADEME, 2010) (Tab. 5).

TABLE 5: GHG balances and LUC factors for several biodiesel pathways.

		(d+i LUC) g eq. CO_2/MJ			
	ADEME, 2010	INRA study (De Cara et al., 2012)			
	(without LUC scenario)	n*	1st quartile	median	3rd quartile
Rapeseed	37.3	79	10	54	90
Sunflower	25.1	10	55	57	59
Soy	21.1	64	56	80	168
Palm	21.8	52	31	55	120
UCOME	8.7				
FAME	8.4				
PVO	31.8	79	10	54	90

n = number of references by pathway, UCOME : used cook oil methyl ester, FAME: fat animal methyl ester, PVO: pure vegetable oil, GHG balance for diesel (ADEME, 2010) = 91.4 g eq. CO_2/MJ (−35% = 59.4, −50% = 45.7).

It can be inferred from Table 5 that the medians values significantly impact GHG balances of biodiesel pathways and may even offset their climate benefits. Thus, adding the median value to the corresponding LCA figure, vegetable oil based biodiesels would not appear to meet the RED sustainability criteria (which over time imposes minimum GHG abatement thresholds of 35, 50 and 60% compared to fossil diesel).

1.4.2 VARIABILITY OF RESULTS

Several studies (De Cara et al., 2012; Malça and Freire, 2011; Plevin et al., 2010) focused on comparing the available environmental assessments of biofuels. Their researches highlight the great variability of results from one assessment to another. For example, emissions associated to biodiesel chains life cycles vary from 15 to 170 gCO_2 eq./MJ. Estimations of both direct and indirect LUC factor (e.g. annualized GHG emissions divided by biofuel energy, expressed in gCO_2 eq./MJ) are among the main sources of variability. However interpreting this variation range of evaluations as the sole reflection of the uncertainty would be a mistake (De Cara et al., 2012). The apparent variability partly reflects the diversity of approaches (LCA vs. economic modeling), definitions and hypothesis in scenario concerning LUC type and original land cover, biofuel pathway, feedstock types and origin, level of mandates, and representation of market mechanisms used in different works. Significant variability is also observed between the results from studies using the same method. When working with LCA methodology, variability between different studies is due to a difference in the choice of approach (attributional or consequential), the choice of system boundaries (well to tank or well to wheel), the choice of the functional unit and the co-products handling (allocation or substitution), while when working with economic models, results depend on the type (general vs. partial equilibrium) and the constructions of models.

A meta-analysis was made by De Cara, et al. (2012) especially in order to quantify the effect of different parameters on assessment of an overall direct plus indirect LUC factor. It shows that results are influenced by:

- the kind of method: LCA lead generally to LUC factor values lower than those provided by economic models;
- the biofuel pathway: all things being equal, bioethanol leads to a LUC factor lower than biodiesel and lignocellulosic ethanol LUC factor seems to be lower than 1st generation ethanol;

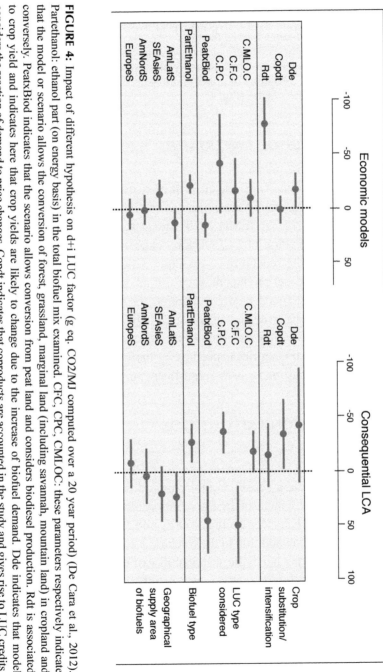

FIGURE 4: Impact of different hypothesis on d+i LUC factor (g eq. CO2/MJ computed over a 20 year period) (De Cara et al., 2012). Partethanol: ethanol part (on energy basis) in the total biofuel mix examined. CFC, CPC, CMLOC: these parameters respectively indicate that the model or scenario allows the conversion of forest, grassland, marginal land (including savannah, mountain land) in cropland and conversely. PeatxBiod indicates that the scenario allows conversion from peat land and considers biodiesel production. Rdt is associated to crop yield and indicates here that crop yields are likely to change due to the increase of biofuel demand. Dde indicates that model considers the reaction of demand to price changes. Copdt indicates that coproducts are accounted in the study and gives rise to LUC credits. AmLatS, SEAsieS, AmNordS, EuropeS described the geographical supply area of biofuel, respectively: Latin America, South East Asia, North America and Europe. Models can consider several areas together.

- the LUC type: when scenario allows conversion of soils with high carbon content (peatlands or forests for example) all things being equal, it predicts significantly higher LUC factors;
- other hypothesis on agricultural yields and elasticity of food demand: scenarios that take into account the yield response and the variation of food demand as a function of prices all things being equal result in lower evaluations of LUC factor (Fig. 4).

1.4.3 SOURCES OF UNCERTAINTY AROUND LUC FACTORS

Many studies on the GHG balances of biodiesel concur in the large uncertainties revolving around the emissions of GHG in the agricultural phase, particularly for N_2O (ADEME, 2010; Bird et al., 2011; Crutzen et al., 2007).

For the estimation of LUC factors, published studies have shown the importance of the reliability of the input data pertaining to both the raw material and biofuel production stages, trade monitoring, supply balances (In Numeri, 2012) and model calibration for sensitive adjustment factors such as hypothesis on display of land used for displaced crop production (De Cara et al., 2012). They also showed the need to improve the monitoring of direct LUC in all countries concerned by biofuel production and trade and the interest of existing tools as Teruti-Lucas survey (AGRESTE, 2010).

1.5 CONCLUSION AND OUTLOOK

At the European level, the development of the biofuels industry, particularly biodiesel, is a sensitive public policy issue. On the one hand, the large-scale deployment of first generation biofuels is quite promising in the sense that it enhances energy security and creates additional opportunities for farmers, in addition to the role that it can play in regional development. On the other hand, the sustainability of biofuels is being questioned since several studies pointed out that the effects of both direct and indirect land use changes triggered by the increase in demand for bioenergy could lead to adverse impacts on the environment.

Life cycle assessment is currently the most recommended methodology to aid decision-making on environmental issues. In this study, we emphasized the need to opt for consequential LCA, in order to encompass both direct and indirect impacts in the evaluation of biofuel chains.

Several approaches have been proposed to quantify the LUCs and assess their environmental effects. However, there is still no consensus on a given method. Indeed, consequential LCA use expert opinions and statistical analysis of historical data to estimate LUC and are often criticized because of the use of simplifying assumptions of market mechanisms, while economic equilibrium models, although they provide strongest estimates of these mechanisms are criticized because their difficulty of use (by nonspecialists) and their often lack transparency. Thus, ensuring an optimal social welfare with biofuels development remains quite difficult.

To generate a more robust assessment of the environmental performance of biofuels, it will be essential to:

- properly assess and isolate land use changes due to biofuels. The use of an economic equilibrium (whether partial orgeneral) model including a land use module with fine spatialresolution and running at a global scale seems to be the mostaccomplished method for achieving this goal;
- provide more accurate estimates of GHG emissions (including CO_2 and N_2O) associated to biofuel feedstock production and LUC, via the use of ecosystem models adapted to local conditions;
- combine economic modeling and LCA, so as to overcome the difficulties related to the tracing of biofuel effects on land use, as observed in other LCA approaches. This will allow us to be more precise when estimating the environmental impacts related to agriculture and land use change. Here, we emphasized that these tools may complement each other. On the one hand, the use of results from economic models in consequential LCA would enhance the quality of iLUC estimation. On the other hand, completing economic models by a life cycle assessment would broaden the range of environmental indicators used to assess biofuels performance, including local impacts such as eutrophication, air quality or toxicity/ecotoxicity.

Parallel, some ways of improvement exist to reduce LUC factors. Measures to increase productivity in agriculture may indeed limit the expansion needed to meet the increased demand related to biofuels and indirect effects of LUC. Improved crop yields (particularly in areas where LUC can have strong impact on GHG emissions such as Latin America or South East Asia) and the energy efficiency of biofuels can reduce the pressure on land and therefore the indirect effects associated with LUC. Genetic improvement could also improve yields as well as reduce the use of inputs.

The technologies that enable to use residues, waste or other feedstock as raw material for biofuel production lie also among ways pointed by EC to reduce LUC and avoid crop displacement and food competition.

Finally, other ways highlighted in the recent Lepage report on biofuels (Lepage et al., 2013) concern the improvement of energy efficiency in transport and the wider use of other renewable energies to contribute to the 10% objective of renewable energy in final consumption of transportation sector in 2020.

NOTE

1.　The first studies are completed and downloadable on ADEME website (http://www2.ademe.fr/servlet/KBaseShow?sort=-1&cid=96&m=3&catid=23698). The last one is still running, results in final validation phase and should be published by ADEME on the same web site during the second quarter of 2013.

REFERENCES

1.　Acquaye AA, Wiedmann T, Feng K, et al. 2011. Identification of carbon hot-spots and quantification of GHG intensities in the biodiesel supply chain using hybrid LCA and structural path analysis, Environ. Sci. Technol. 45: 2471–2478.
2.　ADEME. 2010. Analyses de cycle de vie appliquées aux biocarburants de première génération consommés en France. Étude réalisée pour le compte de l'ADEME par BIO IS 2010, 1–236
3.　AGRESTE. 2004. Enquête TERUTI – série 1992 à 2004. Service de la Statistique et de la Prospective (SSP) du Ministère de l'Agriculture, de l'Alimentation, de la Pêche, de la Ruralité et de l'Aménagement du Territoire.
4.　AGRESTE. 2010. Enquête TERUTI-LUCAS – Nouvelle série 2005 à 2010. Service de la Statistique et de la Prospective (SSP) du Ministère de l'Agriculture, de l'Alimentation, de la Pêche, de la Ruralité et de l'Aménagement du Territoire.
5.　Bauen A, Chudziak C, Vad K, Watson PA. 2010. causal descriptive approach to modelling the GHG emissions associated with the indirect land use impacts of biofuels: Final reports. E4tech, UK department of transport 2010.
6.　Bird N, Cowie A, Cherubini F, Jungmeier G. 2011. Using a life cycle assessment approach to estimate the net greenhouse gas emissions of bioenergy, IEA Bioenergy, Strategic Report, ExCo:2011:03.
7.　Brandão M. 2011. Food, Feed, Fuel, Timber or Carbon Sink? Towards sustainable land-use systems – a consequential life cycle approach, Ph.D. thesis, Centre for Environmental Strategy; Division of Civil, Chemical and Environmental Engineering; Faculty of Engineering and Physical Sciences; University of Surrey.

8. Chakir R, Vermont B. 2013. Étude complémentaire à l'analyse rétrospective des interactions du développement des biocarburants en France avec l'évolution des marches français et mondiaux et les changements d'affectation des sols. Étude réalisée pour le compte de l'ADEME par l'INRA, pp. 1–69.

9. Crutzen PJ, Mosier AR, Winiwarter W. 2007. N2O release from agro-biofuel production negates global warming reduction by replacing fossil fuels, Atmos. Chem. Phys. Discuss. 8: 11191–11205

10. Curran MA. 2013. Life cycle assessment: a review of the methodology and its application to sustainability. Curr. Opin. Chem. Eng. 2: 1–5.

11. De Cara S, Goussebaïle A, Grateau R, et al. 2012. Revue critique des études évaluant l'effet des changements d'affectation des sols sur les bilans environnementaux des biocarburants. Étude réalisée pour le compte de l'ADEME par l'INRA, pp. 1–96.

12. Delucchi M. 2011. A conceptual framework for estimating the climate impacts of land-use change due to energy crop programs. Biomass and Bioenergy 35: 2337–2360.

13. Di Lucia L, Ahlgrem S, Ericsson K. The dilemma of indirect land-use changes in EU biofuel policy – An empirical study of policy-making in the context of scientific uncertainty. Environ. Sci. Policy. 2012: 9–19.

14. Edwards R, Mulligan D, Marelli L. 2010. Indirect land use change from increased biofuels demand: comparison of models and results of marginal biofuels production from different feedstocks, JRC-IE, 2010.

15. Ekvall T, Tillman AM, Molander S. 2005. Normative ethics and methodology for life cycle assessment. J. Clean. Prod. 13: 1225–1234.

16. Ekvall T, Weidema boP. 2004. System boundaries and input data in consequential life cycle inventory analysis. Int. J. LCA 3: 161–171.

17. European Commission. 2009. Directive 2009/28/EC of the European Parliament and of the Council of 23 April 2009 on the promotion of the use of energy from renewable sources and amending and subsequently repealing Directives 2001/77/EC and 2003/30/EC. 2009. Available at: http://eur-lex.europa.eu/LexUriServ/LexUriServ.do?uri=Oj:L:2009:140:0016:0062:en:PDF.

18. European Commission. 2010. Rapport de la Commission sur les changements indirects d'affectation des sols liés aux biocarburants et aux bioliquides. Available at: http://eur-lex.europa.eu/LexUriServ/LexUriServ.do?uri=COM:2010:0811:FIN:FR: PDF.

19. Fargione J, Hill J, Tilman D, Polasky S, Hawthorne P. 2008. Land Clearing and the Biofuel Carbon Debt. Science 319: 1235–1237.

20. Farrel AE, Plevin RJ, Turner BT, Jones AD, O'Hare M, Kammen DM. 2006. O'Hare M, Kammen DM. Ethanol can contribute to energy and environmental goals. Science 311: 506–508.

21. Fritsches UR, Sims R, Monti A. 2010. Direct and indirect land use competition issues for energy crops and their sustainbale production – an overview. Biofuel Bioprod. Biores. 4: 692–704.

22. Gawel E, Ludwig G. 2011. The iLUC dilemma: How to deal with indirect land use changes when governing energy crops?. Land Use Policy 28: 846–856.

23. Gnansounou E, Dauriat A, Villegas J, Panichelli, L. 2009. Life cycle assessment of biofuels : Energy and greenhouse gas balances. Biores. Technol. 100: 4919–4930.

24. Gnansounou E, Panichelli L, Dauriat A, Villegas JD. 2008. Accounting for indirect land use changes in ghg balance of biofuels – Review of current approaches. Lausanne, Switzerland: École polytechnique fédérale de Lausanne, 22 p.

25. Hamelinck C, Koop K, Croezen H, Koper M, Kampman B, Bergsma G. 2008. Technical Specification: Greenhouse Gas Calculator for Biofuels. Utrecht, Netherlands: Ecofys & CE Delft, commissioned by SenterNovem, 109 p.

26. In Numeri. 2012. Analyse rétrospective des interactions du développement des biocarburants en France avec l'évolution des marchés français et mondiaux (productions agricoles, produits transformés et coproduits) et les changements d'affectation des sols. Étude réalisée pour le compte de l'ADEME, pp. 1–128.

27. ISO. 2006a. ISO Norm 14040: Environmental management-life cycle assessment-principles and framework.

28. ISO. 2006b. ISO Norm 14044: Environmental management-life cycle assessment-requirements and guidelines.

29. Kim S, Dale BE. 2011. Indirect land use change for biofuels: Testing predictions and improving analytical methodologies. Biomass and Bioenergy 35: 3235–3240.

30. Kløverpris J, Wenzel H, Nielsen PH. 2008. Life cycle inventory modeling of land use induced by crop consumption. Part 1: conceptual analysis and methodological proposal. Int. J. Life Cycle Assess. 13: 13–21.

31. Laborde D. 2011, Assessing the land use change consequences of European biofuel policies. IFPRI.

32. Laurance WF. 2007. Switch to corn promotes Amazon deforestation. Science 318: 1721.

33. Lechon Y, Cabal H, Sáez R. 2011. Life cycle greenhouse gas emissions impacts of the adoption of the EU Directive on biofuels in Spain. Effect of the import of raw materials and land use changes. Biomass and Bioenergy 35: 2374–2384.

34. Lepage C. 2013. Projet de rapport sur la proposition de directive du Parlement européen et du Conseil modifiant la directive 98/70/CE concernant la qualité de l'essence et des carburants diesel et modifiant la directive 2009/28/CE relative à la promotion de l'utilisation de l'énergie produite à partir de sources renouvelables (COM(2012)0595 – C70337/2012 – 2012/0288(COD)), Commission de l'Environnement, de la santé publique et de la sécurité alimentaire.

35. Malça J, Freire F. 2011. Life-cycle studies of biodiesel in Europe: A review addressing the variability of results and modeling issues. Renew. Sust. Energ. Rev. 15: 338–351.

36. Marelli L, Ramos F, Hiederer R, Koeble E. 2011. Estimate of GHG emissions from global land use change scenarios, JRC Technical notes, JRC-IE.

37. Mellilo JM, Reilly JM, Kicklighter DW, et al. 2009. Indirect Emissions From Biofuels: How important? Science 326: 1397–1399.

38. Nassar AM, Harfuch L, Bachion LC, Moreira MR. Biofuels and land-use changes: searching for the top model. Interface Focus 1: 224–232.

39. Overmars KP, Stehfest E, Ros JPM, Prins AG. 2011. Indirect land use change emissions related to EU biofuel consumption: an analysis based on historical data. Environ. Sci. Policy. 14: 248–257.

40. Plevin RJ, O'Hare M, Jones AD, Torn MS, Gibbs HK. 2010. Greenhouse gas emissions from biofuels' indirect land use change are uncertain but may be much greater than previously estimated. Environ. Sci. Technol. 44: 8015–8021.

41. Reinhard J, Zah R. 2011. Consequential life cycle assessment of the environmental impacts of an increased rapemethylester (RME) production in Switzerland. Biomass and Bioenergy 35: 2361–2373.

42. Reinhard J, Zah R. 2009. Global environmental consequences of increased biodiesel consumption in Switzerland: consequential life cycle assessment. J. Clean. Prod. 17: S46-S56.

43. Searchinger T, Heimlich R, Houghton RA, et al. 2008. Use of US cropland for biofuels increases greenhouse gases through emissions from land-use-change. Science 319: 1238–1240.

44. Searchinger T, Hamburg SP, Melillo J, et al. 2009. Fixing a Critical Climate Accounting Error. Science 326: 527–528.

45. US Environmental Protection Agency. 2009. Draft regulatory Impact analysis: Changes to Renewable Fuel Standard.

46. US Environmental Protection Agency. 2010 Renewable fuel standard program (RFS2) regulatory impact analysis.

47. Van Stappen F, Brose I, Schenkel Y. 2011. Direct and indirect land use changes issues in European sustainability initiatives: State-of-the-art, open issues and future developments. Biomass and Bioenergy 35: 4824–4834.

48. Wang M. 2005. Updated Energy and Greenhouse Gas Emission results of Fuel Ethanol. Center for Transportation Research, Argonne National Laboratory. Available at: http://www.transportation.anl.gov/pdfs/TA/375.pdf.

49. Wang M, Huo H, Arora S. 2011. Methods of dealing with co-productsof biofuels in life-cycle analysis and consequent results within the U.S context. Energ. Policy 39: 5726–5736.

CHAPTER 2

Energy Potential and Greenhouse Gas Emissions from Bioenergy Cropping Systems on Marginally Productive Cropland

MARTY R. SCHMER, KENNETH P. VOGEL, GARY E. VARVEL, RONALD F. FOLLETT, ROBERT B. MITCHELL, AND VIRGINIA L. JIN

2.1 INTRODUCTION

Reduction in greenhouse gas (GHG) emissions from transportation fuels can result in near- and long-term climate benefits [1]. Biofuels are seen as a near-term solution to reduce GHG emissions, reduce U.S. petroleum import requirements, and diversify rural economies. Depending on feedstock source and management practices, greater reliance on biofuels may improve or worsen long-term sustainability of arable land. U.S. farmers have increased corn (*Zea mays* L.) production to meet growing biofuel demand through land expansion, improved management and genetics, increased corn plantings, or by increased continuous corn monocultures [2]–[4]. Productive cropland is finite, and corn expansion on marginally-productive cropland may lead to increased land degradation, including losses in biodiversity and other desirable ecosystem functions [4]–[6]. We define

Energy Potential and Greenhouse Gas Emissions from Bioenergy Cropping Systems on Marginally Productive Cropland. © Schmer MR, Vogel KP, Varvel GE, Follett RF, Mitchell RB, and Jin VL. PLoS ONE, **9,**3 (2014). doi:10.1371/journal.pone.0089501. Originally distributed under the CC0 Universal Public Domain Dedication, http://creativecommons.org/publicdomain/zero/1.0/.

marginal cropland as fields whose crop yields are 25% below the regional average. The use of improved corn hybrids and management practices have increased U.S. grain yields by 50% since the early 1980's [7] with an equivalent increase in non-grain biomass or stover yields. Corn stover availability and expected low feedstock costs make it a likely source for cellulosic biofuel. However, excessive corn stover removal can lead to increased soil erosion and decreased soil organic carbon (SOC) [8] which can negatively affect future grain yields and sustainability. Biofuels from cellulosic feedstocks (e.g. corn stover, dedicated perennial energy grasses) are expected to have lower GHG emissions than conventional gasoline or corn grain ethanol [9]–[13]. Furthermore, dedicated perennial bioenergy crop systems such as switchgrass (*Panicum virgatum* L.) have the ability to significantly increase SOC [14]–[16] while providing substantial biomass quantities for conversion into biofuels under proper management [17], [18].

Long-term evaluations of feedstock production systems and management practices are needed to validate current and projected GHG emissions and energy efficiencies from the transportation sector. In a replicated, multi-year field study located 50 km west of Omaha, NE, we evaluated the potential to produce ethanol on marginal cropland from continuously-grown no-tillage corn with or without corn residue removal (50% stover removal) and from switchgrass harvested at flowering (August) versus a post-killing frost harvest. Our objectives were to compare the effects of long-term management practices including harvest strategies and N fertilizer input intensity on continuous corn grain and switchgrass to determine ethanol production, potential petroleum offsets, and net energy yields. We also present measured SOC changes (0 to 1.5 m) over a nine year period from our biofuel cropping systems to determine how direct SOC changes impact net GHG emissions from biofuels. Furthermore, we evaluate the potential efficiency advantages of co-locating and integrating cellulosic conversion capacity with existing dry mill corn grain ethanol plants.

2.2 MATERIALS AND METHODS

This study is located on the University of Nebraska Agricultural Research and Development Center, Ithaca, Nebraska, USA on a marginal cropland field with Yutan silty clay loam (fine-silty, mixed, superactive, mesic Mollic Hapludalf) and a Tomek silt loam (fine, smectitic, mesic Pachic Argiudoll) soil. Switchgrass plots were established in 1998 and continuous corn plots were initiated in 1999. The study is a randomized complete block design (replications = 3) with split-split plot treatments. Main treatments are two cultivars of switchgrass, 'Trailblazer' and 'Cave-in-Rock', and a glyphosate tolerant corn hybrid. Main treatment plots are 0.3 ha which enables the use of commercial farm equipment. Switchgrass is managed as a bioenergy crop, and corn is managed under no-tillage conditions (no-till farming since 1999). Split-plot treatments are nitrogen (N) fertilizer levels and split-split plots are harvest treatments. Annual N fertilizer rates (2000–2007) were 0 kg N ha^{-1}, 60 kg N ha^{-1}, 120 kg N ha^{-1}, and 180 kg N ha^{-1} as NH_4NO_3, broadcast on the plots at the start of the growing season. The 0 kg N ha^{-1}, 60 kg N ha^{-1}, 120 kg N ha^{-1} fertilizer rates were used on switchgrass [19] while the 60 kg N ha^{-1}, 120 kg N ha^{-1}, and 180 kg N ha^{-1} fertilizer rates were used for corn. Switchgrass harvest treatments were initiated in 2000 and consist of a one-cut harvest either in early August or after a killing frost. Corn stover treatments were initiated in 2000 and are either no stover harvest or stover removal, where the amount of stover removed approximates 50% of the aboveground biomass after corn grain is harvested.

Baseline soil samples were taken in 1998 at the center of each subplot and re-sampled in 2007 at increments of 0–5, 5–10, 10–30, 30–60, 60–90, 90–120, and 120–150 cm depths [15]. Average changes in total SOC (0–1.5 m) from 1998–2007 were used to estimate direct soil C changes. Further management practices and detailed soil property values from this study have been previously reported [15], [20]. Summary of petroleum offsets (GJ ha^{-1}), ethanol production (L ha^{-1}), greenhouse gas (GHG) emissions (g CO_2e MJ^{-1}), net GHG emissions (Mg CO_2e ha^{-1}), and GHG reductions (%) for corn grain, corn grain with stover removal, and switchgrass are presented in Table S1 in File S1.

2.2.1 STATISTICAL ANALYSES

Yield data analyzed were from 2000 to 2007, where 2000 was the initiation of harvest treatments for continuous corn and switchgrass and 2007 was the last year that SOC was measured for this study. Data from switch-grass cultivars were pooled together based on their similar aboveground biomass yields over years and similar changes in SOC [15]. Data were analyzed using a linear mixed model approach with replications considered a random effect. Mean separation tests were conducted using the Tukey-Kramer method. Significance was set at $P \leq 0.05$.

2.2.2 LIFE-CYCLE ASSESSMENT

For energy requirements in the production, conversion, and distribution of corn grain ethanol and cellulosic ethanol, values from the Greenhouse Gases, Regulated Emissions, and Energy Use in Transportation (GREET v. 1.8) [21], Energy and Resources Group Biofuel Analysis Meta-Model (EBAMM) [22], and Biofuel Energy Systems Simulator (BESS) [23] life cycle assessment models were used as well as previous agricultural energy estimates for switchgrass [12]. Energy use in the agricultural phase consisted of agricultural inputs (seed, herbicides, fertilizers, packaging), machinery energy use requirements, material transport, and diesel requirements used in this study. Stover energy requirements from the production phase were from the diesel requirements to bale, load, and stack corn stover and the embodied energy of the farm machinery used. A proportion of the N fertilizer and herbicide requirements were allocated to the amount of stover harvested.

Multiple biorefinery configurations are presented to evaluate different conversion scenarios and how this affects GHG emissions, petroleum offset credits, and net energy yield (NEY) values. Biorefinery scenarios evaluated in this study are: (i) a natural gas (NG) dry mill corn grain ethanol plant with dry distillers grain (DDGS) as a co-product for the corn grain-only harvests [23]–[25], (ii) a co-located dry mill corn grain and cellulosic ethanol plant with combined heat and power (CHP) and DDGS co-product, where corn stover is primarily used to displace dry mill ethanol plant natural gas requirements [25], [26], (iii) and a standalone cellulosic (switchgrass or corn stover) ethanol plant (sequential hydrolysis and fermentation) with CHP capability and electricity export [22], [27]–[29]. Chemical and enzyme production costs and related GHG emissions for corn grain and cellulosic conversion to ethanol were also incorporated [28]. Ethanol recovery for corn grain was estimated to be 0.419 L kg^{-1} [23]. Ethanol recovery for corn stover and switchgrass were based on cell wall composition from harvested biomass samples. Ground aboveground switchgrass samples were scanned using a near-infrared spectrometer to predict cell wall and soluble carbohydrate biomass composition [30]. Ground corn stover samples were analyzed using a near-infrared spectrometer-based calibration equation developed by the National Renewable Energy Laboratory to predict corn stover cell wall composition [31]. Switchgrass and corn stover cell wall conversion to ethanol was based on composition components of glucan, xylose and arabinose [30], [31]. Glucan to ethanol conversion was assumed to be 85.5%, and xylose and arabinose was estimated to have 85% ethanol recovery efficiency [29]. Estimated ethanol recovery for corn stover was 327 L Mg^{-1} which was similar to other findings [29]. For switchgrass, ethanol recovery based on glucan, xylose, and arabinose concentrations was estimated to be 311 L Mg^{-1} and 344 L Mg^{-1} for an August harvest and a post-frost harvest, respectively.

Ethanol plant size capacity was estimated to be 189 million L yr^{-1} for the corn grain-only and cellulosic-only scenarios. For the co-located facility, total plant size was assumed to be 378 million L yr^{-1} capacity. Fossil fuel energy requirements for the conventional corn grain ethanol plant is assumed to be 7.69 MJ L^{-1} for natural gas to power the plant and to dry DGS, 0.59 MJ L^{-1} for corn grain transportation from farm to ethanol plant, 0.67 MJ L^{-1} for electricity purposes, 0.13 MJ L^{-1} to capital depreciation costs, and 0.58 MJ L^{-1} for wastewater processing and effluent restoration [10], [22]. Fossil fuel requirements for the corn grain/cellulosic ethanol plant are feedstock transportation 0.63 MJ L^{-1} for corn stover, 0.59 MJ L^{-1} for corn grain transportation from farm to ethanol plant, 0.44 MJ L^{-1} to capital depreciation costs, and 0.58 MJ L^{-1} for wastewater treatment and processing (Table S2 in File S1). Cellulosic ethanol plant fossil fuel requirements are 0.63 MJ L^{-1} for switchgrass transportation from field to ethanol plant, 0.06 MJ L^{-1} diesel requirements for biomass transport within the ethanol plant grounds, 0.44 MJ L^{-1} to capital depreciation costs, and 0.58 MJ L^{-1} for wastewater processing, effluent restoration, and recovery (Table S2 in File S1).

For the co-located corn grain and cellulosic facility, we assumed (i) power and electrical utilities were shared [26]; (ii) power requirements were supplied mainly from the lignin portion of stover with combined ethanol purification from the starch and cellulosic ethanol conversion pathways [26]; and (iii) extra stover biomass would be required in addition to the lignin to meet steam requirements. A co-location facility would require additional unprocessed bales to be used in addition to lignin which lowered the amount of ethanol being generated from stover at a co-located facility compared to a standalone cellulosic facility that uses stover as their primary feedstock (Table S1 in File S1). Electricity would be imported from the grid in this scenario and DDGS exported as the only co-product. Recent analysis [29] of converting cellulose to ethanol has estimated a higher internal electrical demand than previously assumed [26]; suggesting electricity export under this configuration would be unlikely. The value of DDGS as animal feed would likely preclude its use in meeting power requirement in a co-located facility. We based our total biomass energy requirement on the lignin concentration in stover and the expected biomass energy use requirements to power a co-located ethanol plant [25]. Estimated biomass requirements were 11 MJ L^{-1} ethanol and embodied energy value of 16.5 MJ kg^{-1} (low heating value) for stover biomass.

Net energy yield (NEY) values (renewable output energy – fossil fuel input energy) were calculated for each feedstock and conversion scenario. Output energy was calculated from ethanol output plus co-product credits. Co-product credit for DDGS is 4.13 MJ L^{-1} for the corn grain-only ethanol plant and the co-located corn grain/cellulosic ethanol plant [32]. Electricity co-product credit for standalone cellulosic ethanol was estimated to be 1.68 MJ L^{-1} [29]. Petroleum offsets (GJ ha^{-1}) were calculated in a similar fashion as NEY with total ethanol production (MJ ha^{-1}) along with petroleum displacement from co-products minus petroleum inputs consumed in the production, conversion, and distribution phase (Tables S1 and S3 in File S1). Petroleum offsets were calculated as the difference between ethanol output and petroleum inputs from the agricultural, conversion, and distribution phase (Table S1 in File S1). Petroleum requirements for each cropping system were calculated from input requirements from this study and derived values from the EBAMM model [22]. For input requirements without defined petroleum usage, we used the default parameter in

EBAMM that estimates U.S. average petroleum consumption at 40% for input source. Petroleum offset credits associated with corn grain ethanol co-products were estimated to be 0.71 MJ L^{-1} while credits for corn stover and switchgrass cellulosic ethanol co-products (standalone facility) were 0.12 MJ L^{-1} (Table S3 in File S1). Petroleum offset credits were calculated from GREET (v 1.8).

2.2.3 GREENHOUSE GAS EMISSIONS

Greenhouse gas offsets associated with the production of corn grain and cellulosic ethanol were modeled from the EBAMM and BESS models [22], [23]. Agricultural GHG emissions were based on fuel use, fertilizer use, herbicide use, farm machinery requirements, and changes in SOC. Direct land use change by treatment plot can either be a GHG source or a GHG sink depending on SOC changes from this study [15]. Co-product GHG credits for DDGS or electricity export were derived from the BESS [23] and GREET (v. 1.8) models [21]. Co-product GHG credits for DDGS was −347 g CO_2e L^{-1} ethanol and −304 g CO_2e L^{-1} ethanol for cellulosic electricity export (Table S4 in File S1). Indirect land use changes for corn grain ethanol or switchgrass were not estimated in this analysis. GHG offsets were calculated on both an energy and areal basis (Table S1 in File S1).

Greenhouse gas emissions from N fertilizer were evaluated from the embodied energy requirements and subsequent nitrous oxide (N_2O) emissions (Table S4 in File S1). Direct and indirect nitrous oxide emissions were calculated in this study using Tier 1 Intergovernmental Panel on Climate Change calculations. Greenhouse gas emission values for the agricultural phase are included in Table S4 in File S1 and for the conversion and distribution phase in Table S5 in File S1. For the agricultural phase, total GHG emissions were calculated from the production of fertilizers, herbicides, diesel requirements, drying costs for corn grain, and the embodied energy in farm machinery minus direct soil C changes occurring for the study period (Table S4 in File S1). GHG emissions were reported on an energy basis, areal basis, and the difference between ethanol and conventional gasoline (Table S1 in File S1). For net GHG emissions (Mg CO_2e ha^{-1}), calculations were based on GHG intensity values (g CO_2e

MJ^{-1}) multiplied by biofuel production (MJ ha^{-1}) for each cropping system. GHG reductions (Table S1 in File S1) were calculated as the percent difference from conventional gasoline as reported by the California Air Resource Board (99.1 g CO_2 MJ^{-1}) [33].

2.3 RESULTS AND DISCUSSION

Harvest and N fertilizer management treatments affected grain and biomass yields in both crops over eight growing seasons (Fig. 1A). Switchgrass harvested after a killing frost had 27% to 60% greater biomass yields compared with an August harvest under similar fertilization rates. Highest harvested biomass yields (mean = 11.5 Mg ha^{-1} yr^{-1}) were from fertilized (120 kg N ha^{-1}) switchgrass harvested after a killing frost while continuous corn showed similar grain and stover yields [factorial analysis of variance (ANOVA), P = 0.72] under the highest N fertilizer levels (180 kg N ha^{-1}) (Fig. 1A).

Potential ethanol yields varied from 2050 to 2774 L ha^{-1} yr^{-1} for corn grain-only harvests while those for corn grain with stover removal ranged from 2862 to 3826 L ethanol ha^{-1} yr^{-1} (Fig. 1B). Ethanol contribution from corn stover ranged from 820 to 998 L ha^{-1} yr^{-1} when stover is converted at a standalone cellulosic plant (Fig. 1B). Separate ethanol facilities showed slightly higher potential ethanol yields (L ha^{-1}) than at a co-located facility (Table S1 in File S1) because a larger portion of corn stover biomass was required to meet thermal power requirements at a co-located facility (SI text in File S1). Unfertilized switchgrass had potential ethanol yield values similar to corn stover. Switchgrass under optimal management practices had 17% higher biomass yields than the highest yielding corn with stover removal treatment. Potential ethanol yield for switchgrass, however, was similar (factorial ANOVA, P>0.05) to corn with stover removal (Fig. 1B) due to lower cellulosic ethanol recovery efficiency than exists for corn grain ethanol conversion efficiency. Switchgrass ethanol conversion efficiency from this study was based on updated biochemical conversion processes [29] using known cell wall characteristics [30] that result in lower conversion rates than previous estimates [12], [18].

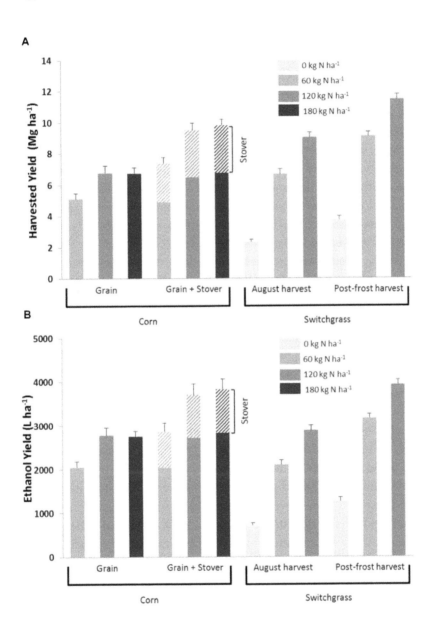

FIGURE 1: Harvested mean annual yield ± standard error (A) and ethanol energy ± SEM (B) for no-till continuous corn (grain-only harvest or grain and stover harvest) and switchgrass (August harvest or Post-frost harvest) under variable nitrogen rates on marginally-productive rainfed cropland for 2000–2007 (n = 3 replicate corn system plots and 6 replicate switchgrass plots).

Net energy yield (NEY) (renewable output energy minus fossil fuel input energy) and GHG emission intensity (grams of CO_2 equivalents per megajoule of fuel, or g CO_2e MJ^{-1}) are considered the two most important metrics in estimating fossil fuel replacement and GHG mitigation for biofuels [34]. Switchgrass harvested after a killing frost (120 kg N ha^{-1}) and the co-located grain and stover conversion pathway (120 kg N ha^{-1} and 180 kg N ha^{-1} treatments) had the highest overall NEY values (Fig. 2). Net energy yields for continuous corn were higher at a co-located facility because stover biomass and lignin replaced natural gas for thermal energy (Fig. 2). Ethanol conversion of corn grain and stover at separate facilities was intermediate in NEY while traditional corn grain-only natural gas (NG) dry mill ethanol plants had the lowest NEY values for the continuous corn systems. Delaying switchgrass harvest from late summer to after a killing frost resulted in significant improvement in NEY and potential ethanol output under similar N rates. Unfertilized switchgrass had similar NEY values compared with corn grain processed at a NG dry mill ethanol plant (factorial ANOVA, $P = 0.12$) while fertilized switchgrass harvested after a killing frost had higher NEY values (factorial ANOVA, $P<0.0001$) than NG dry mill corn grain ethanol plants (Fig. 2).

Both the continuous corn and switchgrass systems showed significant petroleum offset (ethanol output minus petroleum inputs) capability, with the intensified bioenergy cropping systems having the highest petroleum offsets (Fig. 3). Petroleum use varied by cropping system in the agricultural phase with continuous corn systems having higher overall petroleum requirements than switchgrass. Petroleum requirements (mainly diesel fuel) to harvest corn stover are small relative to corn grain harvest as a result of low harvested stover yields. Lowest petroleum offsets for continuous corn systems were from stover harvests at a separate dedicated cellulosic facility (Table S1 in File S1). Corn grain-only harvests offset less petroleum compared with grain and stover at separate ethanol facilities under similar fertilizer rates (factorial ANOVA, $P<0.01$). Management practices in switchgrass resulted in the largest variation in petroleum offset credits (Fig. 3B). Petroleum offsets (GJ ha^{-1}) were positively associated with NEY values [$-1.81+0.84$ (Petroleum offset); ($P<0.0001$); ($R^2 = 0.76$)], indicating that bioenergy cropping systems with large NEY values will likely result in higher petroleum displacement.

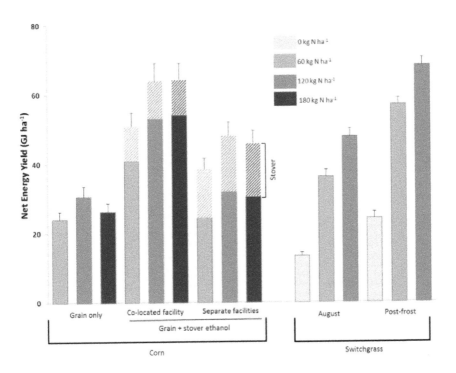

FIGURE 2: Net energy yield ± standard error for no-till continuous corn (grain-only or grain and stover harvest) and switchgrass (August harvest or post-frost harvest) under variable nitrogen rates on marginally-productive cropland (n = 3 replicate corn system plots and 6 replicate switchgrass plots). Conversion processes evaluated include corn grain-only harvest at a natural gas (NG) dry mill, corn grain with stover harvest at a co-located facility (lignin portion of stover used as primary energy source for grain and cellulose conversion), corn grain with stover harvest at separate ethanol facilities (NG dry mill and a cellulosic ethanol plant), and switchgrass (cellulosic ethanol plant).

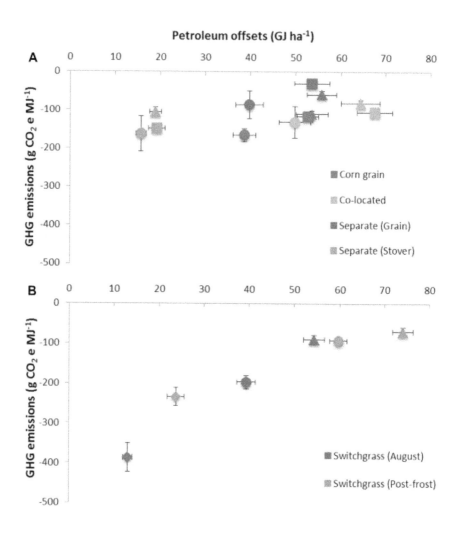

FIGURE 3: Petroleum offsets compared with GHG emissions (g CO_2e MJ^{-1} ethanol) for continuous corn and switchgrass grown on marginally-productive cropland (n = 3 replicate corn system plots and 6 replicate switchgrass plots). (A) Continuous corn values represent harvest method (stover harvested or retained) and ethanol conversion pathway (co-located facility or at a separate ethanol facilities). (B) Switchgrass values are based on harvest date and N fertilizer rate. Fertilizer rates are 0 kg N ha^{-1} (♦), 60 kg N ha^{-1} (•), 120 kg N ha^{-1}, and 180 kg N ha^{-1} (•). Error bars indicate standard errors of the mean.

All bioenergy cropping systems evaluated in our study had SOC sequestration rates exceeding 7.3 Mg CO_2 yr^{-1} (Table S4 in File S1), with over 50% of SOC sequestration occurring below the 0.3 m soil depth [15]. Soil organic C increased even with corn stover removal, indicating that removal rates were sustainable in terms of SOC and grain yield for this time period. No-tillage continuous corn systems have lower stover retention requirements to maintain SOC than continuous corn with tillage or corn-soybean (*Glycine max* (L.) Merr.) rotations [8]. Consequently, all conversion pathways had negative GHG emission values as a result of SOC sequestration offsetting GHG emissions from the production, harvest, conversion and distribution phases for corn grain ethanol and cellulosic ethanol. For switchgrass, SOC storage values were similar to other findings within the same ecoregion [16] and a long-term Conservation Reserve Program grassland [35]. Measured SOC storage from the continuous corn systems (Table S4 in File S1) were significantly higher than modeled SOC storage estimates from this region [36]. Corn grain grown with low N rates (60 kg ha^{-1}) had GHG intensity values similar to continuous corn under optimum N rates (120 kg ha^{-1}) but resulted in lower ethanol yields and lower petroleum offset potential (Fig. 3A). Lowest GHG emission intensity values on an energy basis (g CO_2e MJ^{-1}) were from unfertilized switchgrass (Table S1 in File S1) due to lower ethanol yields, lower agricultural energy emissions, and similar SOC storage compared with the other biofuel cropping systems. For switchgrass, management practices that resulted in the lowest GHG emission on an energy basis resulted in the lowest petroleum offset potential (Fig. 3B). Direct N_2O emissions (Table S4 in File S1) were estimated using Intergovernmental Panel on Climate Change methodology and are in agreement with study site N_2O flux measurements from a later time series which indicated N rate as the major contributor to N_2O emissions [37]. When evaluating GHG emissions on a per unit area basis (g CO_2e ha^{-1}), unfertilized switchgrass and corn grain-only systems showed similar results with the more intensified cropping systems (Table S1 in File S1).

Both switchgrass and continuous corn with stover removal produced similar ethanol potential, NEY values, petroleum offsets, and GHG emissions but overall values and metric efficiencies were dependent on management practices and downstream conversion scenarios. Dedicated perennial grass systems used for bioenergy will need to have similar or greater yield potential than existing annual crops for widespread adoption to meet renewable energy demands and provide similar economic returns to producers. We have previously shown that switchgrass ethanol yields were comparable with regional corn grain ethanol yields [12]. Here we demonstrate that when switchgrass is optimally managed, ethanol potential is similar to a continuous corn cropping system with stover removal and exceeds ethanol yield for corn grain-only systems on marginally-productive cropland. Furthermore, breeding improvements for bioenergy specific switchgrass cultivars have shown higher yield potential than cultivars evaluated here [38].

Coupling sustainable agricultural residue harvests with dedicated energy crops improves land-use efficiency and reduces biomass constraints for a mature cellulosic biofuel industry. Recent analysis has shown that sufficient land exists in the U.S. Corn Belt to support a cellulosic ethanol industry without impacting productive cropland [18], [39], [40]. The effect of dedicated energy crops and corn grain on indirect land use change varies significantly based on the assumptions and models used [13], [41], [42] but bioenergy crops grown on marginally-productive cropland will have less impact on indirect land use change than bioenergy crops grown on more productive cropland. Likewise, model assumptions underlying direct SOC sequestration will impact system evaluations of GHG emissions and mitigation. Measured SOC sequestration values presented here were based on production years evaluated and were not extrapolated beyond this time-frame. Extrapolating SOC values from this time-frame to a 30-yr time horizon or 100-yr time horizon is still larger than current life cycle assessment assumptions on SOC sequestration potential of switchgrass or no-till corn [12], [42], [43]. This highlights the importance of accounting for direct SOC changes at depth to accurately estimate GHG emissions for biofuels under both marginal and productive cropland. Further long term evaluation of management practices (e.g. tillage, stover removal) on SOC sequestration potential for corn grain systems under irrigated conditions on productive cropland is warranted [44].

A multi-feedstock, landscape approach minimizes economic and environmental risks in meeting feedstock demands for cellulosic ethanol production by providing sufficient feedstock availability while maintaining ecosystem services. A co-located cellulosic biorefinery is expected to have economic advantages by reducing capital costs requirements for cellulosic conversion and through sharing of infrastructure costs. In this study, we used corn stover as the feedstock for the co-located cellulosic biorefinery but the benefits will apply to other cellulosic feedstocks. A co-located facility can increase NEY values by decreasing natural gas use for thermal energy, but current and forecasted U.S. natural gas prices [45] may affect large scale adoption of co-location unless there are incentives for displacing fossil energy in existing NG dry mill ethanol plants [46]. Integrating cellulosic refining capacity with existing corn grain ethanol plants can improve the sustainability of first generation biofuels and enable the implementation of cellulosic biofuels into the U.S. transportation sector.

REFERENCES

1. Unger N, Bond TC, Wang JS, Koch DM, Menon S, et al. (2010) Attribution of climate forcing to economic sectors. Proc Natl Acad Sci USA 107(8): 3382–3387. doi: 10.1073/pnas.0906548107

2. Wallander S, Claassen R, Nickerson C (2011) The ethanol decade: An expansion of U.S. corn production, 2000–09. USDA-ERS Economic Information Bulletin No. (EIB-79).

3. Claassen R, Carriazo F, Cooper J, Hellerstein D, Ueda K (2011) Grassland to cropland conversion in the northern plains: The role of crop insurance, commodity, and disaster programs Economic Research Report No. (ERR-120).

4. Wright CK, Wimberly MC (2013) Recent land use change in the western corn belt threatens grasslands and wetlands. Proc Natl Acad Sci USA 110(10): 4134–4139. doi: 10.1073/pnas.1215404110

5. Wiens J, Fargione J, Hill J (2011) Biofuels and biodiversity. Ecol Appl 21(4): 1085–1095. doi: 10.1890/09-0673.1

6. Donner SD, Kucharik CJ (2008) Corn-based ethanol production compromises goal of reducing nitrogen export by the Mississippi river. Proc Natl Acad Sci USA 105(11): 4513–4518. doi: 10.1073/pnas.0708300105

7. USDA-NASS. 2013 U.S. Corn Yields. Available: http://www.nass.usda.gov/Charts_and_Maps/Field_Crops/cornyld.asp.

8. Wilhelm WW, Johnson JMF, Karlen DL, Lightle DT (2007) Corn stover to sustain soil organic carbon further constrains biomass supply. Agron J 99(6): 1665–1667. doi: 10.2134/agronj2007.0150

9. Adler PR, Del Grosso SJ, Parton WJ (2007) Life-cycle assessment of net greenhouse-gas flux for bioenergy cropping systems. Ecol Appl 17(3): 675–691. doi: 10.1890/05-2018

10. Farrell AE, Plevin RJ, Turner BT, Jones AD, O'Hare M, et al. (2006) Ethanol can contribute to energy and environmental goals. Science 312(5781): 506–508. doi: 10.1126/science.1121416

11. Sheehan J, Aden A, Paustian K, Killian K, Brenner J, et al. (2003) Energy and environmental aspects of using corn stover for fuel ethanol. J Indust Ecol 7(3–4): 117–146. doi: 10.1162/108819803323059433

12. Schmer MR, Vogel KP, Mitchell RB, Perrin RK (2008) Net energy of cellulosic ethanol from switchgrass. Proc Natl Acad Sci USA 105(2): 464–469. doi: 10.1073/pnas.0704767105

13. Wang M, Han J, Dunn JB, Cai H, Elgowainy A (2012) Well-to-wheels energy use and greenhouse gas emissions of ethanol from corn, sugarcane and cellulosic biomass for US use. Environ Res Letters 7(4): 045905. doi: 10.1088/1748-9326/7/4/045905

14. Frank AB, Berdahl JD, Hanson JD, Liebig MA, Johnson HA (2004) Biomass and carbon partitioning in switchgrass. Crop Sci 44: 1391–1396. doi: 10.2135/cropsci2004.1391

15. Follett RF, Vogel KP, Varvel GE, Mitchell RB, Kimble J (2012) Soil carbon sequestration by switchgrass and no-till maize grown for bioenergy. BioEnerg Res 5: 866–875. doi: 10.1007/s12155-012-9198-y

16. Liebig MA, Schmer MR, Vogel KP, Mitchell RB (2008) Soil carbon storage by switchgrass grown for bioenergy. BioEnerg Res 1(3–4): 215–222. doi: 10.1007/s12155-008-9019-5

17. DOE US (2011) U.S. billion-ton update: Biomass supply for bioenergy and bioproducts industry. ORNL/TM-2011/224.

18. Gelfand I, Sahajpal R, Zhang X, Izaurralde R, Gross KL, et al. (2013) Sustainable bioenergy production from marginal lands in the US Midwest. Nature 493: 514–517. doi: 10.1038/nature11811

19. Vogel KP, Brejda JJ, Walters DT, Buxton DR (2002) Switchgrass biomass production in the Midwest USA: Harvest and nitrogen management. Agron J 94: 413–420. doi: 10.2134/agronj2002.0413

20. Varvel GE, Vogel KP, Mitchell RB, Follett RF, Kimble J (2008) Comparison of corn and switchgrass on marginal soils for bioenergy. Biomass Bioenergy 32(1): 18–21. doi: 10.1016/j.biombioe.2007.07.003

21. Greenhouse Gases, Regulated Emissions, and Energy Use in Transportation (GREET) (2012) Argonne National Laboratory. Available: http://greet.es.anl.gov/.

22. Renewable and Applicable Energy Laboratory (2007) Energy and resources group biofuel analysis meta-model. Available: http://rael.berkeley.edu/sites/default/files/EBAMM/.

23. Liska AJ, Yang HS, Bremer VR, Erickson G, Klopfenstein T, et al.. (2008) BESS: Biofuel energy systems simulator; life-cycle energy and emissions analysis model for corn-ethanol biofuel (v. 2008.3.0).

24. Liska AJ, Yang HS, Bremer VR, Klopfenstein TJ, Walters DT, et al. (2009) Improvements in life cycle energy efficiency and greenhouse gas emissions of corn-ethanol. J Ind Ecol 13(1): 58–74. doi: 10.1111/j.1530-9290.2008.00105.x

25. Wang M, Wu M, Huo H (2007) Life-cycle energy and greenhouse gas emission impacts of different corn ethanol plant types. Environ Res Letters 2(2): 024001. doi: 10.1088/1748-9326/2/2/024001

26. Wallace R, Ibsen K, McAloon A, Yee W (2005) Feasibility study for co-locating and integrating ethanol production plants from corn starch and lignocellulosic feedstocks. USDOE Rep No. NREP/TP-510-37092.

27. Spatari S, Bagley DM, MacLean HL (2010) Life cycle evaluation of emerging lignocellulosic ethanol conversion technologies. Bioresource Technol 101(2): 654–667. doi: 10.1016/j.biortech.2009.08.067

28. MacLean HL, Spatari S (2009) The contribution of enzymes and process chemicals to the life cycle of ethanol. Environ Res Letters 4(1): 014001. doi: 10.1088/1748-9326/4/1/014001

29. Humbird D, Davis R, Tao L, Kinchin C, Hsu D, et al.. (2011) Process Design and Economics for Biochemical Conversion of Lignocellulosic Biomass to Ethanol. USDOE (Department of Energy NREL/TP-5100-47764, Golden, CO), pp 136.

30. Vogel K, Dien BS, Jung HG, Casler MD, Masterson SD, et al. (2011) Quantifying actual and theoretical ethanol yields for switchgrass strains using NIRS analyses. BioEnerg Res 4(2): 96–110. doi: 10.1007/s12155-010-9104-4

31. Templeton DW, Sluiter AD, Hayward TK, Hames BR, Thomas SR (2009) Assessing corn stover composition and sources of variability via NIRS. Cellulose 16: 621–639. doi: 10.1007/s10570-009-9325-x

32. Graboski MS (2002) Fossil energy use in the manufacture of corn ethanol. National Corn Growers Association

33. California Air Resources Board (2013) Low carbon fuel standard. Available: http://www.arb.ca.gov/Fuels/Lcfs/Lcfs.htm.

34. Liska AJ, Cassman KG (2008) Towards standardization of life-cycle metrics for biofuels: Greenhouse gas emissions mitigation and net energy yield. J Biobased Materials and Bioenergy 2(3): 187–203. doi: 10.1166/jbmb.2008.402

35. Gelfand I, Zenone T, Jasrotia P, Chen J, Hamilton SK, et al. (2011) Carbon debt of conservation reserve program (CRP) grasslands converted to bioenergy production. Proc Natl Acad Sci USA 108(33): 13864–13869. doi: 10.1073/pnas.1017277108

36. Davis SC, Parton WJ, Del Grosso SJ, Keough C, Marx E, et al. (2012) Impact of second-generation biofuel agriculture on greenhouse-gas emissions in the corn-growing regions of the US. Front Ecol Environ 10: 69–74. doi: 10.1890/110003

37. Jin VL, Varvel GE, Wienhold BJ, Schmer MR, Mitchell RB, et al.. (2011) Field emissions of greenhouse gases from contrasting biofuel feedstock production systems under different N fertilization rates. ASA-CSSA-SSSA International Annual Meetings. Oct 16–19, San Antonio TX.

38. Vogel KP, Mitchell RB (2008) Heterosis in switchgrass: Biomass yield in swards. Crop Sci 48(6): 2159–2164. doi: 10.2135/cropsci2008.02.0117

39. Mitchell RB, Vogel KP, Uden DR (2012) The feasibility of switchgrass for biofuel production. Biofuels 3(1): 47–59. doi: 10.4155/bfs.11.153

40. Uden DR, Mitchell RB, Allen CR, Guan Q, McCoy T (2013) The feasibility of producing adequate feedstock for year-round cellulosic ethanol production in an intensive agricultural fuelshed. BioEnerg Res 6(3): 930–938. doi: 10.1007/s12155-013-9311-x

41. Searchinger T, Heimlich R, Houghton RA, Dong F, Elobeid A, et al. (2008) Use of U.S. croplands for biofuels increases greenhouse gases through emissions from land-use change. Science 319(5867): 1238–1240. doi: 10.1126/science.1151861

42. Dunn J, Mueller S, Kwon H, Wang M (2013) Land-use change and greenhouse gas emissions from corn and cellulosic ethanol. Biotechnology for Biofuels 6(1): 51. doi: 10.1186/1754-6834-6-51
43. Fargione J, Hill J, Tilman D, Polasky S, Hawthorne P (2008) Land clearing and the biofuel carbon debt. Science 319(5867): 1235–1238. doi: 10.1126/science.1152747
44. Follett RF, Jantalia CP, Halvorson AD (2013) Soil carbon dynamics for irrigated corn under two tillage systems. Soil Sci Soc Am J 77: 951–963. doi: 10.2136/sssaj2012.0413
45. Energy Information Agency (2013) Annual energy outlook for 2013 with projections to 2040. DOE/EIA-0383. doi: 10.1787/9789264179233-en
46. Plevin RJ, Mueller S (2008) The effect of CO2 regulations on the cost of corn ethanol production. Environ Res Letters 3(2): 024003. doi: 10.1088/1748-9326/3/2/024003

There are several supplemental files that are not available in this version of the article. To view this additional information, please use the citation on the first page of this chapter.

CHAPTER 3

Integration of Farm Fossil Fuel Use with Local Scale Assessments of Biofuel Feedstock Production in Canada

J. A. DYER, R. L. DESJARDINS, B. G. MCCONKEY, S. KULSHRESHTHA, AND X. P. C. VERGÉ

3.1 INTRODUCTION

The viability of Canadian biofuel industries will depend on farm energy consumption rates and the CO_2 emissions from fossil fuel use for feedstock crops. The types of biofuels that are under development in Canada include biodiesel, grain ethanol, cellulosic ethanol and biomass. Each of these fuels relies on a distinct class of feedstock crops and in each case the most suitable crop is also dependent on geographic location. For example, the feedstock for biodiesel is canola in Western Canada and soybeans in Eastern Canada (Dyer et al., 2010a). For grain ethanol, the feedstock choices are corn in the east and wheat in the west (Klein and LeRoy, 2007). Cellulosic ethanol is still under development in Canada.

Technological changes in ethanol manufacturing can bring about different intensities of land use and require different land capabilities. Cellulosic ethanol and biomass can make use of land not capable of growing grains, and can exploit part of the straw from annual field crops (Dyer et al., 2011a). As a result, impacts on other land use activities with which feedstock crops compete also depend on the particular feedstock involved in the interaction and the capability of the land. Impacts on the overall sustainability of agriculture are minimal when management practices fit the local environment (Vergé et al., 2011). Therefore, to understand the different comparative advantages and impacts among regions, each landscape requires its own assessment.

Two main principles must guide biofuel industries. The first is that they must produce more energy than the fossil energy used for their production. The second is that they must displace more Greenhouse Gas (GHG) emissions than are released during their production (Dyer and Desjardins, 2009; Klein and LeRoy, 2007). Biofuels appeal to governments for the potential to create economic opportunities in rural areas (Klein and LeRoy, 2007). Due to transport costs, feedstock crops are best grown on land that is close to facilities for processing them into biofuel. Thus, it is important to have objective criteria for determining which communities and regions are the most suitable locations for those processing plants. In addition, sustainable feedstock production requires that local suitability be established (Dyer et al., 2011a; Vergé et al., 2011). To date, a comprehensive farm energy analysis has not been done at a local scale in Canada.

The main goal of this chapter was to determine the geographic distribution of farm energy terms within each province of Canada. Due to their small sizes and limited role in Canadian agriculture, the four Atlantic Provinces were treated as one combined province. A secondary goal of this chapter was to demonstrate how much the farm energy budget contributes to the GHG emissions budget of the agricultural sector through fossil CO_2 emissions at a provincial scale. Using area based intensity, a simple demonstration was also provided of how these data could provide a baseline comparison for the fossil CO_2 emitted from growing a grain ethanol feedstock compared to current types of farms. These goals were achieved through the integration of existing models and databases, rather than by analysis of new data collected specifically for this purpose.

3.2 BACKGROUND

The feedstock for biofuels has raised several land use questions (GAO, 2009; Malcolm and Aillery, 2009). These include: How much land will biofuel feedstock production require in order for biofuels to make an appreciable contribution to energy supply? What agricultural products would be displaced to accommodate this production? How will food supply be threatened by feedstock production? How much will meat production and livestock industries be displaced by feedstock? In large part, most of these general land use policy questions have been addressed in Canada and elsewhere. However, there have been some shortcomings of these analyses.

One of these gaps is the failure by many studies to account for carbon dioxide (CO_2) emissions caused by fossil fuel use in the feedstock production, and in agriculture, generally. One of the reasons for this gap is that under the United Nations Framework Convention on Climate Change, emissions from fossil fuels used for agriculture are reported as part of the energy sector, rather than under the agriculture sector. Although smaller in magnitude than both the methane (CH_4) and nitrous oxide (N_2O) emissions reported for agriculture, farm energy-related CO_2 emissions are an important component of the sector's GHG emissions budget, largely because it is manageable (Dyer and Desjardins, 2009). For example, reduced tillage practices which diminish fossil fuel CO_2 emissions from farm machinery (Dyer and Desjardins, 2003a), as well as conserving soil carbon, can be the difference in whether a particular feedstock or its biofuel are energy-positive or a sink for GHGs.

Without taking all forms of fossil energy use in agriculture into account, the GHG emissions budget for crop production is incomplete. In addition to farm field operations, the fossil fuel CO_2 emissions include agro-chemical manufacturing, equipment manufacturing, fuels for grain drying or heating farm buildings, gasoline, and electricity for lighting or cooling (Dyer and Desjardins, 2009). However, farm field operations are the most complex term and have the greatest degree of interaction with land features and crop choices. Fossil fuel consumption for farm field work has been computed using the Farm Field work and Fossil Fuel Energy and Emissions (F4E2) model (Dyer and Desjardins,

2003b; 2005). Because of their dominant role in defining regional differences in fossil fuel energy and CO_2 emissions, farm field operations have already been assessed in more detail than other farm energy terms (Dyer et al., 2010b).

3.3 METHODOLOGY

3.3.1 SELECTING THE SPATIAL SCALE

Since decision making in the biofuel industries is limited by spatial scale, assessing the most appropriate scale was the first task undertaken in this analysis. Disaggregation of the Canadian farm energy budget to the provinces can exploit agricultural statistics available at two spatial scales. The first scale is the Census Agricultural Regions (CAR) (Statistics Canada, 2007), while the second scale is at the Soil Landscapes of Canada (SLC) (AAFC, 2011). Due to its association with agricultural census records, the geographic scale chosen for distributing farm energy use in this chapter was the CAR system which divides Canada into 55 regions (with each of the Atlantic Provinces treated as a single CAR). In spite of the soil and land variables available for SLCs, some difficult assumptions are needed to disaggregate some data to this scale. In addition to this uncertainty, the large number of spatial units in Canada at the SLC scale (nearly 4,000 units having agriculture) made presentation on the basis of SLCs impractical for this chapter.

The CARs are identified in this chapter by numbers that start from 1 in each province. In the Atlantic Provinces, with each province treated as one CAR. Hence, CAR numbers 1, 2, 3, and 4 represent New Brunswick, Prince Edward Island, Nova Scotia and Newfoundland, respectively. With the agricultural regions of Canada being spread out largely east to west, it was not practical to display the boundaries on a single page map. So, a website location, rather than a printed map, was provided in this chapter. To view the CAR sizes and locations in each province, visit: http://www.statcan.gc.ca/ca-ra2011/110006-eng.htm.

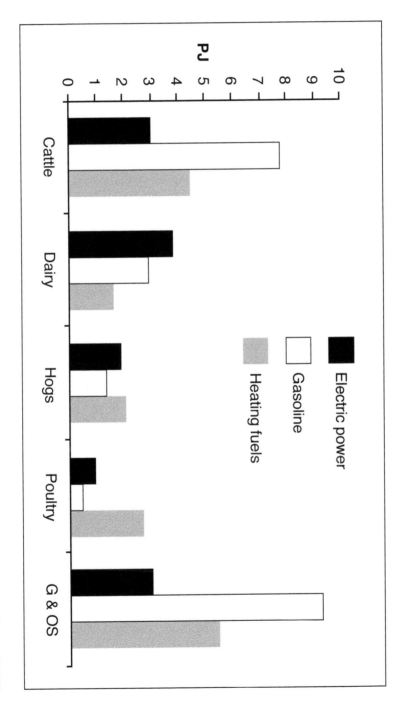

FIGURE 1: National consumption of three types of energy by five farm types identified in the 1996 Farm Energy Use Survey (FEUS) of Canada.

3.3.2 FARM ENERGY BUDGET

The six terms in the farm energy budget adopted for this analysis were those defined by Dyer and Desjardins (2009). All of these terms reflect operational and/or financial decisions made by farmers. For example, the energy costs of transporting products from farm gate to market that are paid for by the processer or marketer, rather than the farmer, were excluded. These terms involved several different types of fossil fuel. Based on the analytical methodologies required for spatial disaggregation, these six terms were separated into three groups. The diesel fuel used in farm field work (Dyer and Desjardins, 2003b; 2005) and the coal required to manufacture and supply farm machinery (Dyer and Desjardins, 2006a) were the first group because they were both quantified with the F4E2 model.

The fossil energy to supply chemical fertilizers and pesticide sprays was determined from a direct conversion of the weight of consumption of these chemicals (Dyer and Desjardins, 2007). Since nitrogen fertilizers are the most energy-intensive chemical inputs to manufacture, and have available sales records in Canada, this conversion was based on the natural gas to manufacture just nitrogen fertilizer. The energy conversion rate of 71.3 GJ/t{N} derived from Nagy (2001) as an average for five census years from 1981 to 2001 was used in this chapter. Although this conversion was for just nitrogen supply, it was indexed to include other farm chemicals, mainly phosphate and potash fertilizers.

The third group includes electrical power, gasoline and heating fuels. All three terms in this group had to be determined empirically since there was little basis for modeling these terms. While to some extent diesel is increasingly being used for farm owned transport vehicles, in 1996 the F4E2 model accounted for all but a small percentage (Dyer and Desjardins, 2003b; 2005) of the farm-purchased diesel fuel for farm field work. Only one percent of this diesel fuel was for household use in 1996 (Tremblay, 2000). This suggests that pick-up trucks, the sort of vehicle that would be used for both light haul farm transport and family business, were not typically diesel powered in 1996. Therefore, gasoline, rather than diesel, was likely the main fuel used for farm owned transport vehicles in 1996, the baseline year for the farm energy budget described by Dyer and Desjardins (2009). There was, therefore, no justification for including any diesel fuel in the third group of energy terms. In keeping with the conditions of the farm energy budget described above, any diesel fuel consumed by commercial trucks used for hauling grain and livestock to market or processing were not considered in this analysis.

Electrical power was a partial exception to the need for empirical determination because of a semi-empirical index of the CO_2 emissions from this term based on farm types (Dyer and Desjardins, 2006b). This index demonstrated the correlation, at least for this energy term, between energy consumption and farm types, particularly among livestock farms. Application of this index for this analysis was unnecessary because in this case livestock populations are only needed to distribute a known quantity of electrical energy among provinces and regions (CARs).

The most comprehensive source of farm energy use information in Canada is the 1996 Farm Energy Use Survey (FEUS) of Canada (Tremblay, 2000). The FEUS provided commodity-specific estimates for the three energy terms for which detailed modeling algorithms were not available. Given this empirical source, for example, it did not matter whether all gasoline was burned in farm owned transport vehicles or whether all such vehicles were powered by gasoline. What mattered was that the FEUS provided an empirical quantity of gasoline that had to be disaggregated regionally. The remaining term in the Canadian farm energy budget was a combination of three fuels, including furnace-oil, liquid propane (LPG) and natural gas, which was defined by Dyer and Desjardins (2009) as heating fuels.

Due to confidentiality constraints, the FEUS data were not directly available at the farm level. The FEUS, however, did allow energy type data to be grouped by farm type, but only for Canada as a whole. While energy types were also grouped by provinces in the FEUS, this breakdown could not be linked to farm type uses. The FEUS also gave the consumption of diesel fuel in Canadian agriculture which was used to verify the F4E2 model (Dyer and Desjardins, 2003b). The quantities for the farm energy terms extracted from the FEUS, shown in Figure 1, illustrate the range in energy quantities that had to be disaggregated for these three energy types. These energy data were adjusted for the shares of these fuels that were used in farm households instead of farm use. These household share adjustments were only provided by fuel type, however, and not for farm type (Tremblay, 2000).

Although the purpose of the data in Figure 1 was not to compare farm types, these energy quantities still reflect both the different sizes and energy intensities of these farming systems in Canada. Grain and oilseed farms accounted for 35% of the consumption of these three energy terms. The range of total live weights in Canada for beef, dairy, hogs and poultry of 5.7, 1.1, 0.8 and 0.2 Mt, respectively, during 2001 (Vergé et al., 2012) was wider than the range in uses of these three energy types among the four livestock industries seen in Figure 1. Hence, while beef production used the largest share of this energy of any of the livestock industries, beef farms were the least intensive user on a live weight basis. Similarly, poultry, the smallest livestock industry and lowest user of these energy terms, was the most intensive user of these three types of energy.

3.3.3 LAND USE

In defining the GHG emission budgets for each of the Canada's four dominant types of livestock production, dairy, beef, pork and poultry, Vergé et al. (2007; 2008; 2009 a,b) took into consideration the land base on which the feed grains (including oilseed meal) and forage that support livestock are grown. Vergé et al. (2007) recognized that the carbon footprint for each livestock industry must include the land base that supports the crops in the livestock diet. Subsequently, the total area involved in Canadian livestock production was defined as the Livestock Crop Complex (LCC). The LCC was based on an array of crops that defined the diets of all four livestock types, including barley, grain corn, soybean meal, feed quality wheat, oats, canola meal, dry peas, seeded pasture, alfalfa, grass hay and silage corn.

The Canadian Economic and Emissions Model for Agriculture (CEEMA) was developed to estimate the spatial distribution and magnitude of GHG emissions generated by the agriculture sector (Kulshreshtha et al., 2000). Because the spatial unit of CEEMA was the CAR, this model was well suited for the analysis described in this chapter. CEEMA is composed of records of crop areas, yields, nitrogen fertilizer rates and related GHG emissions during 2001 for all field crops in each CAR. Almost 1,900 of these crop records were distributed over 55 CARs in CEEMA. While crop records identify the CAR in which they lie and define the areas of all crops within each CAR, the actual locations of crops described in the respective records within the CAR are not specified. Another limitation of the CEEMA was that these crop records were generated from analysis of optimal economic land uses for 2001 (Horner et al., 1992; Kulshreshtha et al., 2000), rather than from actual crop statistics.

The variables that determine differences among the CARs are related primarily to land use differences and farm level decisions. These variables include the selections of crops, particularly those crops that feed livestock. The CEEMA crop records do not contain soil type data. Livestock populations at the CAR scale were also not available for this analysis to preserve the confidentiality of the farmers surveyed at that scale. The variables required for assessing farm energy at the CAR scale will be discussed in more detail below.

Estimates of GHG emissions from Canada's four main livestock industries were integrated with the CEEMA. The area of each crop that was in the LCC from each CAR in each province was determined as part of a previous application of CEEMA (Dyer et al., 2011b). That study disaggregated the LCC to each crop record describing crops in the diet of Canada's four main livestock types. Some feedstock-food-livestock interactions on a national or provincial scale in Canada were analyzed in that study. It also used the CEEMA database to separate Canadian farmland into land that supported livestock and land available for other crops. However, Dyer et al. (2011b) did not separate these emissions by livestock type. Farm energy consumption and fossil fuel CO_2 emissions for farm field work have been disaggregated at a provincial scale (Dyer et al., 2010b). But no other farm energy terms have been disaggregated at a scale that allows the full farm energy budget to be quantified in the CARs.

3.3.4 FARM ENERGY AND LIVESTOCK DISTRIBUTIONS

For the three energy terms that can only be treated empirically, electric power, gasoline and heating fuels, the FEUS provided the only link to farm types. Because of the availability of provincial livestock population data from the Canadian agricultural census, this disaggregation can be done directly at the provincial scale. Grain and oilseed production, which was defined as a farm type in the FEUS, accounted for part of each of these three energy terms. Therefore, provincial summaries of areas in these crops were also involved in the disaggregation process.

TABLE 1: The provincial 2001 energy quantities for the three energy terms and the five farm types identified at a national scale in the FEUS1.

	Beef	Dairy	Hogs	Poultry	G&OS[2]
			PJ		
		Electric power			
British Columbia	0.15	0.25	0.02	0.14	0.00
Alberta	1.71	0.26	0.32	0.10	0.75
Saskatchewan	0.59	0.10	0.22	0.04	1.60
Manitoba	0.33	0.13	0.48	0.07	0.35
Ontario	0.27	1.28	0.58	0.40	0.06
Quebec	0.12	1.20	0.70	0.26	0.02
Atlantic	0.03	0.20	0.06	0.06	0.01
Canada	3.20	3.42	2.40	1.07	2.78
		Gasoline[3]			
British Columbia	0.39	0.19	0.02	0.07	0.01
Alberta	4.43	0.20	0.23	0.05	2.33
Saskatchewan	1.52	0.08	0.15	0.02	4.95
Manitoba	0.85	0.10	0.34	0.03	1.08
Ontario	0.70	0.98	0.41	0.19	0.18
Quebec	0.32	0.92	0.49	0.12	0.05
Atlantic	0.09	0.15	0.04	0.03	0.02
Canada	8.30	2.63	1.68	0.51	8.61
	Heating fuel[4]				
British Columbia	0.22	0.11	0.02	0.40	0.01
Alberta	2.53	0.11	0.35	0.28	1.37
Saskatchewan	0.87	0.04	0.24	0.10	2.92
Manitoba	0.49	0.06	0.53	0.20	0.64
Ontario	0.40	0.54	0.64	1.12	0.10
Quebec	0.18	0.51	0.77	0.72	0.03
Atlantic	0.05	0.08	0.07	0.18	0.01
Canada	4.75	1.45	2.62	2.99	5.08

[1]*1996 Farm Energy Use Survey for Canada* [2]*grains and oilseed farms* [3]*gasoline purchased by fann operators for famt-owned vehicles.* [4]*includes furnace-oil, liquid propane (LPG) and natural gas*

These farm type links meant that disaggregation of these energy terms to the CAR scale could be achieved through correlation with livestock populations and crop areas. The underlying assumption was that most farm animals are located near their feed sources. This assumption was required because information on where in the provinces farm animals are actually housed was not available for this analysis (Tremblay, 2000). This limitation only affected the three empirical energy terms, including electric power needs, heating fuels and gasoline for farm transport. The farm field work and the two input supply terms can be linked directly to the CARs through CEEMA, as well as to the provinces.

Provincial estimates had to be generated for all three energy terms taken directly from the FEUS. To achieve this, the relative distribution of energy quantities across the provinces was determined for each farm type identified in the FEUS. To quantify each livestock farming system, the inter-provincial distribution was determined on the basis of the total weight of all live animals in all age-gender categories in the livestock type. The provincial live weight was calculated from the average live weight (W) of each age-gender category (k) of each livestock type (a) and the number of head (H) in each age-gender category and livestock type. The amount of energy from each energy term for each of the livestock systems from the FEUS ($E_{FEUS,a}$) was disaggregated to the provincial energy quantity (E_{prov}) by the respective shares of live weight in each province (prov), as follows.

$$E_{prov,a} = E_{FEUS,a} \times (\textstyle\sum k\ Wk,a \times Hk, prov,a) / (\textstyle\sum Canada \textstyle\sum k\ Wk,a \times Hk, prov,a) \quad (1)$$

The disaggregation of these energy terms for the farms that produce grains and oilseeds to the provinces was similar to Equation 1. The difference was that live weights (W × H) were replaced by the provincial crop areas in this farming system. The areas of each grain and oilseed crop were summed over the crop records of grain or oilseed areas in the CEEMA database. The first sum was for the crop records in each CAR to determine CAR area totals. The provincial totals for each type of grain or oilseed crop were then estimated from the sum of all areas in that crop type over all CARs in each province. This summing process was only applied to the actual grains and oilseeds crops. So rather than correlate the entire area in these crops with the energy terms, differences between these area totals and the areas of these annual crops in the LCC were used. Dyer et al. (2011b) defined these areas as the Non-Livestock Residual areas (NLR). The provincial quantities for the three energy terms and the five farm types shown in Figure 1 are given in Table 1.

A simpler computational sequence was used for the two energy terms derived from the F4E2 model and the energy term for chemical inputs. This was possible because the data for calculating these terms could be taken directly from the crop records of the CEEMA database. The main input variable from CEEMA for the F4E2 calculations was crop areas, whereas total chemical nitrogen applications were available in all CEEMA crop records for the chemical input supply energy term. Because these two energy terms were calculated on each crop record, they could be summed directly from the CEEMA database. While the calculations for grains and oilseeds used only the records for those crops designated as grains and oilseeds, calculations for these three terms used all crop records associated with the LCC or NLR. The F4E2 model took into account whether the crops were annual grains or perennial forages, along with the yields of each crop (Dyer and Desjardins, 2005).

The analysis for this chapter did not disaggregate provincial livestock populations directly into the CARs. Instead, it was the LCC areas defined by these populations that were disaggregated at this scale. Like the NLR area summations, only crop records for those crops that were in each respective livestock diet were summed within the CARs, rather than the areas from all crop records in the CEEMA database. The basis for identifying these crop records was the set of provincial LCC calculations for each livestock type provided by Vergé et al. (2012).

Since the FEUS data were collected in 1996 and the CEEMA data were derived from the 2001 agricultural census, the energy quantities in Figure 1 had to be indexed from 1996 to 2001. This was done by factoring the 1996 energy terms by the ratio of the respective size of each farm system from the 2001 census records to the size of the same farm system in the 1996 census records. Updating from 1996 to 2001 was done at the same time as the farm type energy quantities from the FEUS were disaggregated to the provinces, as shown in Table 1. The different farm types required different definitions of size. For the four livestock farm types, these provincial size ratios were of total livestock weights from the two years, whereas for grain and oilseed farm areas (NLR), these provincial ratios were of total crop production (planted areas times yields) from the two census years.

3.3.5 AREA ALLOCATION TO EACH CAR

The allocation of LCC areas (A) to each CAR for each livestock type was determined by the aggregate share of all feed crops in the provincial LCC in that CAR. Crop areas from the crop records were converted to area totals in each CAR for each of the 12 LCC crops (listed above) that were common to both the CEEMA database and to the four LCCs (Vergé et al., 2012). The total LCC areas in the crop records (Dyer et al., 2011b) were integrated to the respective CARs for each livestock type. The allocation to livestock types was based on the share of each of the four LCCs in each province, which were derived from the diet of each livestock population (Vergé et al., 2012).

For ruminant livestock, the allocation of provincial energy quantities to the CARs required a means of equating the dietary contribution of roughages with that of feed grains. For ruminants, 1.8 kg of roughages provide the same nutrient energy as 1 kg of feed grains (IFAS, 1998; Neel, 2012; Schoenian, 2011). Using this ratio, the forages in the respective LCC areas were converted to the equivalent feed grains on the basis of crop production estimates derived from the 2001 census crop yields. This general relationship also applies to pulses and oilseed meals, but ignores the protein contributions from those feeds. This relationship is altered slightly for corn silage which provides only 42% by weight of the nutrient value of other roughages (Miller and Morrison, 1950). Rangeland was excluded because there were no available data for farm energy consumption associated with this form of land use. Very little energy would be consumed to manage rangeland because no fertilizer or chemical inputs are used and, normally, there are no farm field operations.

In addition to the different nutritional values, the bulk yield differences between grains (g) and roughages (r) also account for the importance of these two crop group areas in each LCC. For each CAR the total LCC area

(A_{CAR}) was the result of the two areas ($A_{CAR,g}$ and $A_{CAR,r}$). Each area was weighted by the average total production weights for the crop group (F) within each provincial LCC and 1.8 (the nutritional value ratio for g and r). This weighted area total was calculated for each CAR as follows.

$$A_{CAR} = ((A_{CAR,g} \times F_g) + (A_{CAR,r} \times F_r/1.8)) / (F_g + (F_r / 1.8)) \qquad (2)$$

These LCC area calculations at the CAR level were integrated over each province as follows.

$$A_{prov} = \Sigma_{CAR} A_{CAR} \qquad (3)$$

Each energy term (E) from the FEUS for each province (Equation 1) was then disaggregated from the province to the CAR level as follows.

$$E_{CAR} = E_{prov} \times (A_{CAR} / A_{prov}) \qquad (4)$$

Dyer et al. (2011b) found that occasionally the amounts of some crops were too low to meet the dietary needs of the provincial livestock populations. Because of these crop deficits, production from the surplus provinces had to be transported to the deficit provinces. Due to the reduction of $A_{CAR,r}$ by 1/1.8 and the occasional accumulation of these provincial crop deficits and surpluses, A_{CAR} was an indexed area estimate which did not equal the actual total LCC area for the CAR. Without reducing $A_{CAR,r}$ by 55%, A_{prov} would have the same difference with the provincial LCC area total (prior to these deficit corrections) as each A_{CAR} would have with the CAR total of the LCC area. Thus, using the CAR to province area ratios of these two weighted area estimates to disaggregate provincial energy terms does not result in any unnecessary distortion of the CAR energy estimates compared to the CAR-province ratios of uncorrected LCC areas.

The usefulness of disaggregating to the CAR scale depends on the sensitivity of the farm energy terms to land use parameters. Since the goal of this chapter was to determine the spatial distribution of the farm energy budget, a sensitivity analysis based on purely management-based range tests such as those described by Dyer and Desjardins (2003a) would not adequately demonstrate the sensitivity of farm energy terms to the factors that determine the spatial distribution of farm energy use at the CAR scale. This was because the only available spatial parameter at the sub-provincial CAR scale was the array of crop areas from CEEMA. Instead, the spatial sensitivity was equated to the variance of energy estimates across CARs in each province. Such sensitivity would reveal the impacts of local crop choice decisions on the consumption of different energy types.

3.4 RESULTS AND DISCUSSION

3.4.1 FARM ENERGY BUDGET AT THE CAR SCALE

The basic output from the analysis described in this chapter was the set of disaggregated farm energy terms at the CAR scale. Due to the extent of these data, they are presented in appendices, rather than as tabular results in the main body of the chapter. Some care is needed in the numbering system in these appendices since the website maps for two provinces use a different CAR numbering system than was used in CEEMA. For Manitoba, CEEMA CAR number 1 includes the online map numbers 1, 2 and 3; CEEMA number 2 includes the online map numbers 4, 5 and 6; and CEEMA number 5 includes the online map numbers 9 and 10. For CEEMA numbers 3, 4 and 6, the online map numbers are 7, 8 and 11, respectively. The online map number 12 was not used in CEEMA. To be consistent with the online CAR base map, the 10 CEEMA CARs for Ontario were combined into 5 CARs in the two Appendices.

The data presented in Appendix A are preliminary to the general (non-commodity-specific) farm energy budget in Appendix B. They resulted from the need to use farm types to disaggregate the FEUS data. The data presented in Appendix B are the intended output or primary goal of this chapter. These data represent all six terms in the energy budget described by Dyer and Desjardins (2009). The data for the three energy terms extracted from the FEUS in Appendix B were derived by integrating the data in Appendix A over the five farm types. Although it is difficult to extract any trends from these data arrays by inspection that could not otherwise be seen from provincial scale tables, these two appendices make the data at the CAR scale available for future regional investigations in farm energy use in Canada.

TABLE 2: Provincial estimates of the six energy tenns of the Canadian farn energy budget duing 2001.

Provinces	Farm field work	Machinery supply	Chemical inputs	Electric power	Gasoline[1]	Heating fuel[2]
			PJ			
British Columbia	1.1	0.6	1.0	0.6	0.7	0.8
Alberta	19.4	11.1	34.8	3.1	7.2	4.6
Saskatchewan	31.7	18.2	35.3	2.5	6.7	4.2
Manitoba	10.6	6.1	21.3	1.4	2.4	1.9
Ontario	8.6	4.9	11.1	2.6	2.5	2.8
Quebec	4.7	2.7	6.5	2.3	1.9	2.2
Atlantic	0.9	0.5	1.4	0.4	0.3	0.4
Canada	76.9	44.1	111.4	12.9	21.7	16.9

[1]*gasoline purchased by fann operators for farn-owned vehicles.* [2]*includes furnace-oil, liquid propane (LPG) and natural gas*

3.4.2 PROVINCIAL FARM ENERGY

Table 2 presents a re-integration of the spatially detailed data in Appendix B from the CAR to provincial scale. Even given the limited spatial detail of this table, it still puts all terms of the Canadian farm energy budget into one source, based on one integrated methodology. Not surprisingly, given its large crop area, Saskatchewan was the biggest consumer of all forms of farm energy in Canada. This was most evident in the farm machinery-related terms, which likely reflects the extensive grains and oilseeds farming system in that province. The two coastal regions (British Columbia and the Atlantic Provinces), as well as Quebec and Ontario, contribute much less to the farm energy budget than the three Prairie Provinces, simply because of the much smaller areas in agricultural use. Although fertilizers (and other farm chemicals) are the largest cause of energy consumption, the farm machinery-related terms combined are 9% higher, nationally, than the chemical inputs. The three FEUS-based terms, to which so much attention was devoted in this chapter, account for only 20% of the national farm energy budget.

3.4.3 ASSESSING SENSITIVITY THROUGH SPATIAL VARIANCE

The spatial variance assessments of the spatial data in this chapter are shown in Tables 3 and 4. The statistic used to compare spatial variance was the coefficient of variation (CV) of the CAR energy values within each province. Being the ratio of standard deviations to their respective means, the CVs give a normalized, and thus a comparable, measure of spatial variability. In order to avoid the CVs being affected by the sizes of the CARs, the data in the two appendices were converted to energy intensities using areas of arable land extracted from the CEEMA crop records (discussed in more detail below). To illustrate, if the disaggregated energy intensities are evenly dispersed across all CARs in the province, then the crop records in the CEEMA database would have no impact on the distribution of energy consumption. Evenly dispersed energy quantities across all CARs would also result in no variance among the CARs and a provincial CV of zero.

Table 3 presents the CVs for the data presented in Appendix A, while Table 4 presents the CVs for Appendix B. In Table 3, only one set of CV estimates was needed for all three energy terms since there was no source of spatial variation associated with these energy terms prior to disaggregation to the CARs. For the pork, poultry and grains and oilseeds farm types, the two coastal provinces had the highest CVs in Table 3. Manitoba had the lowest CVs for these three farm types, which were also the lowest CVs in Table 3. For dairy, Quebec had the lowest CV, while for beef, the lowest

CV was in Alberta. The poultry industry had the highest spatial variation, followed by grains and oilseeds, while dairy had the lowest overall spatial variation. Spatial variation for pork and poultry was lowest in the Manitoba. The spatial variations for pork and poultry were generally higher than for beef and dairy. On average, the Prairies had lower CVs than the other provinces. All of the CVs in Table 3 were higher than zero and there were appreciable differences among these CVs. Hence, the crops that drive these five farming systems were not evenly distributed among the CARs.

TABLE 3: Provincial Coefficients of Variation (CV) for the disaggregation of energy use by five farm types from the FEUS[1] to the CARs[2] during 2001.

Provinces	Beef	Dairy	Pork	Poultry	G&OS[3]
			cv[4]		
British Columbia	0.31	0.30	0.84	1.27	0.90
Alberta	0.25	0.18	0.34	0.33	0.39
Saskatchewan	0.35	0.25	0.30	0.26	0.27
Manitoba	0.39	0.27	0.19	0.18	0.16
Ontario	0.53	0.20	0.38	0.40	0.13
Quebec	0.30	0.13	0.65	0.65	0.38
Atlantic	0.39	0.32	0.83	1.01	0.82

[1]*Farm Energy Use Survey,* [2]*census agricultural regions of Canada,* [3]*grains and oilseed farns,* [4]*these CV estimates represent all three energy terms from the FEUS.*

Whereas there were no spatial differences among the three energy terms from the FEUS when they were separated by their farm types (Table 3), integrating over those five farm types in Table 4 created some differences among these three energy terms. Since farm field work and machinery supply were connected to each other through the F4E2 model, and had the same spatial variations, only the farm field work CVs were shown in Table 4. Farm field work, electric power and gasoline use all had similar CVs which were all lower than the CVs for heating fuels and chemical inputs. The higher CVs for heating fuels likely reflect the combining of three fuel types into one term.

Manitoba had the lowest average CV over the five energy terms in Table 4. British Columbia had the highest CVs for all energy terms except chemical inputs, which were highest in Saskatchewan. The Atlantic Provinces and then Quebec had the next highest CVs after British Columbia. The CV for electric power in Ontario was so low that it suggested almost no spatial differences for this term in Ontario. There was not as much within-province variation among the energy terms (Table 4) as among the farm types (Table 3) that determined the spatial variations for three of those terms. The CVs in Table 4 still display an appreciable amount of within-province spatial variation, however.

The more hilly and ecologically-varied terrain in the coastal provinces may account for some of the spatial variance in British Columbia and the Atlantic Provinces compared to the prairies. However, the agricultural areas in the Prairie Provinces, particularly Saskatchewan, are greater than in the other provinces, and have a greater range in latitude, and hence climate, which would result in higher spatial variation among the CARs. In spite of the relatively low CVs in some cases, Tables 3 and 4 still suggest that the data presented in the two appendices can provide some guidance on where in each province farm energy use would be the highest or the lowest for each energy term.

TABLE 4: Provincial Coefficients of Variation (CV) for the disaggregation of the six energy terns in the Canadian farm energy budget to the CARs[1] during 2001.

	Farm field work	Chemical inputs	Electrical power	Gasoline	Heating fuel
Provinces			CV		
British Columbia	0.32	0.33	0.47	0.34	0.70
Alberta	0.20	0.26	0.12	0.12	0.11
Saskatchewan	0.07	0.44	0.16	0.17	0.16
Manitoba	0.08	0.22	0.14	0.13	0.12
Ontario	0.11	0.16	0.02	0.13	0.14
Quebec	0.25	0.11	0.16	0.08	0.34
Atlantic	0.25	0.28	0.23	0.20	0.45

[1]*census agricultural regions of Canada*

3.4.4 FOSSIL CO_2 EMISSIONS FROM FARM ENERGY USE

To satisfy the secondary goal of this chapter the farm energy budget presented in Table 2 was converted to fossil CO_2 emissions. With the variety of energy types that are used in Canadian agriculture, a different conversion was required for each of the six energy terms. For the diesel fuel for field work, coal to manufacture steel for farm machinery and gasoline, the conversion factors were 70.7, 86.2 and 68.0 Gg{CO_2}/PJ (Neitzert et al., 2005). Based on a summary of fertilizers manufacturing energy dynamics by Nagy (2001), Dyer and Desjardins (2007) used 57.9 Gg{CO_2}/PJ as the conversion factor for fossil CO_2 emissions from fertilizer supply. Even though the chemical input supply energy computations were driven by just nitrogen applications, this conversion took into account all three fertilizers, not just nitrogen, since all three fertilizers were included in this energy term. Reasoning that a very small additional share of the input energy was devoted to the supply of pesticides, which were not included in the calculations from Nagy (2001), Dyer and Desjardins (2009) defined this CO_2 emissions term as chemical inputs, rather than fertilizer supply.

Because heating fuel includes three separate fossil fuels, CO_2 emission rates had to be determined for each farm type in the same way as energy consumption rates for heating fuel were determined. This was done by converting the set of fuel and farm type estimates for this energy term and converting them to CO_2 emissions, using 59.8, 61.0 and 67.7 Gg{CO_2}/PJ, for LPG, natural-gas and furnace-oil (Neitzert et al., 2005). The conversion factor for each fuel and farm type was the ratio of these CO_2 emissions and the previously discussed energy consumption amounts. The blended factors had only minor variation among the provinces, however, ranging from 61.8 Gg{CO_2}/PJ for Saskatchewan to 66.6 Gg{CO_2}/PJ for the Atlantic Provinces. Therefore, the average heating fuel conversion factor for Canada, 64.1 Gg{CO_2}/PJ, was used for all provinces in Table 5.

Since they were interested in a national farm energy budget, Dyer and Desjardins (2009) used a single average conversion factor for CO_2 emissions for the consumption of electric power. Their factor allowed for 22% of Canadian electricity generation being from coal-fired plants. However, there are great differences among provinces in the dependence of coal-based generation (NRCan, 2005), ranging from 96% in Alberta to 0% in

Quebec. Because of the goal of provincial disaggregation of all farm fossil CO_2 emissions to provinces in this chapter, the conversion factor for each province was computed separately using the provincial percent of coal generation from each province. The resulting conversion factors were 41.4, 264.8, 209.6, 2.8, 44.1, 0.0 and 162.4 Gg{CO_2}/PJ, respectively, for British Columbia, Alberta, Saskatchewan, Manitoba, Ontario, Quebec and the Atlantic Provinces.

TABLE 5: Provincial fossil CO_2 emissions from the six terms of the Canadian farm energy budget during 2001.

Provinces	Farm field work	Machinery supply	Chemical inputs	Electric power	Gasoline[1]	Heating Fuel[2]
			$GgCO_2$			
British Columbia	79	55	61	23	46	49
Alberta	1,372	961	2,014	832	492	295
Saskatchewan	2,238	1,567	2,044	533	457	258
Manitoba	749	524	1,231	4	163	122
Ontario	605	423	643	114	167	183
Quebec	332	233	378	0	130	145
Atlantic	60	42	81	59	23	26
Canada	5,435	3,805	6,451	1,566	1,478	1,078

[1]*gasoline purchased by farm operators for farm-owned vehicles.* [2]*includes furnace-oil, liquid propane (LPG) and natural gas.*

Like Table 2, the provincial differences in Table 5 reflect the range in sizes of the agriculture sector in the provinces. Saskatchewan accounted 36% of the fossil CO_2 emissions, while the three Prairie Provinces accounted for 80%. The two coastal provinces only accounted for 4%. While fertilizer supply was the largest energy term, the two terms related to farm field work exceeded fertilizer supply as a CO_2 emitter by 50%. Heating fuels had the lowest emissions, both for Canada and for all of the provinces. The three terms from the FEUS emitted only 21% of the fossil CO_2 from Canadian agriculture. The greatest variation among provinces was from the electric power term, due to the provincial differences in the use of coal for generating power. Heating fuels showed the least variation among provinces.

With a few minor adjustments to methodology, the basic energy budget described in this chapter (prior to spatial disaggregation) was very similar to the national energy budget presented by Dyer and Desjardins (2009). Therefore, the total emissions for Canada in Table 5 can be compared to the CO_2 totals for 2001 in that paper. Dyer and Desjardins (2009) showed higher CO_2 emissions for gasoline and heating fuels than this chapter because that analysis included several horticultural farm systems that were not included in the CEEMA database. Electric power CO_2 emissions were higher in this chapter than the emissions from this term by Dyer and Desjardins (2009). This was due to the decision to use province-specific energy to CO_2 conversions for electric power generation in this chapter, which captured the greater dependence on coal in the provinces with the largest agriculture sectors. The national CO_2 emissions estimate for three energy terms that could be computed directly from the CEEMA crop records in this chapter were all equal to the 2001 estimates reported by Dyer and Desjardins (2009).

3.4.5 ENERGY USE AND CO2 EMISSION INTENSITIES

The farm fossil fuel associated with feedstock production would depend on the specific type of feedstock crop to be produced. The data in Appendix B provide a set of baseline data against which the fossil fuel required for a specific feedstock crop choice would have to be compared. These data represent the mean quantities of farm energy used either for food or livestock feed production in each CAR. These mean energy quantities, summarized by province in Table 2 and converted to CO_2 emissions in Table 5, were also converted to the area based intensities shown in Figure 2 using the crop areas presented in Table 6. These areas include annual crops and seeded perennial forages summarized from the CEEMA crop records to the CAR scale. The CARs in Table 6 are numbered in the same sequence that was used in the two appendices. Because the areas in unseeded pasture and other marginal lands account for almost no farm energy use in Canada, they were not included in Table 6. These data can be used with Appendix B to calculate the intensity of energy use in each CAR (and were used in Tables 3 and 4). Over 80% of the arable land in Canada is in the three Prairie Provinces, and almost half of Canada's farmland is in Saskatchewan.

TABLE 6: Areas in annual crops and seeded perennial forages distributed over the 55 Census Agricultural Regions (CAR) of Canada during 200 L

CAR #	British Columbia	Alberta	Saskatchewan	Manitoba	Ontario	Quebec	Atlantic Provinces
				ha, 000			
1	11	766	1,252	1,413	822	182	113
2	22	1,036	1,351	750	1,083	94	128
3	27	940	2,198	665	373	45	90
4	17	1,993	733	781	624	73	6
5	65	1,107	2,108	397	444	102	
6	2	974	2,196	505		97	
7	57	1,513	1,342			116	
8	185		1,569			84	
9			1,849			210	
10						472	
11						219	
Total	386	8,329	14,598	4,511	3,347	1,695	337

Figure 2 integrates the six energy terms in each province. Figure 2a shows the mean energy use per ha while Figure 2b shows the mean CO_2 emissions per ha. Although the distribution of CO_2 emissions resembles the distribution of energy uses across the provinces, there are slight differences because of the different farm type mixes and fuel types associated with those farm types among the provinces. Saskatchewan had the lowest energy use and CO_2 emission intensities because that province has the lowest share of its arable land devoted to livestock feed.

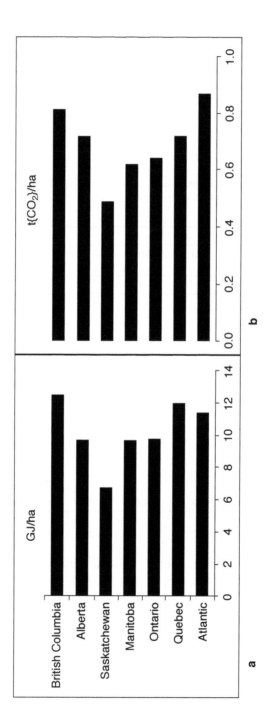

FIGURE 2: Area based intensity of on-farm energy use (a) and fossil CO_2 emissions (b) from all farm types in each province of Canada in 2001.

The following example illustrates how to reconcile biofuel feedstock production with farm fuel use and fossil CO_2 emissions. Using their 2009 methodology, Dyer and Desjardins (2007) described theoretical CO_2 emission budgets for a wheat farm in Saskatchewan and a dairy farm in Ontario. From the perspective of carbon footprint, the simulated wheat farm would be similar to a farm growing grain as a feedstock for ethanol. Based solely on fossil CO_2 emissions, the emission intensity for the ethanol feedstock crop was only 0.26 t/ha, compared to the mean intensity of 0.49 t/ha for all farm types in Saskatchewan in Figure 2b. This result suggests that diverting farmland to grow ethanol feedstock might actually lower the average on-farm fossil CO_2 emissions in Saskatchewan.

In Ontario, the simulated dairy farm emission intensity described by Dyer and Desjardins (2007) was 0.62 t/ha, compared to the 0.64 t/ha for all farm types in the province. This close agreement reflects the high share of Ontario farmland that is devoted to livestock production, much of which is dairy. These comparisons ignore CO_2 emissions from the soil, as well as the other types of GHG. A similar comparison would also be possible for the energy required to grow other biofuel feedstock crops based on data presented in this chapter. Since it is often debatable what the correct land base should be when comparing per ha intensities of different farm types, Figure 2 should be viewed with caution. Farm land has a wide range of capabilities and intensities of use. Therefore the efficiency of food or feedstock production is not necessarily determined by land use intensity.

3.5 SUMMARY AND CONCLUSIONS

Quantifying the local impacts from land use changes driven by expanding markets for biofuel was a major focus in this chapter. The degree of spatial detail for the complete farm energy budget presented here is unprecedented in Canada. The sensitivity analysis technique for farm energy demonstrated by Dyer and Desjardins (2003a) could be used to assess scenarios for the growth of biofuel industries. While this has been done for livestock to biofuel feedstock interactions in Canada (Dyer et al., 2011c), more detailed spatial resolution for such scenario or sensitivity analysis is required. With the three Prairie Provinces accounting for 80% of both the arable land and overall farm energy use in Canada (Tables 2 and 6), the ability to assess the energy consumption patterns in this region in more spatial detail than at the provincial scale is especially important. The procedure described in this chapter disaggregated all terms to the CAR scale before re-integrating to the provincial scale. Because of this quantitative link with the CARs, and its computational flexibility, this procedure is ideally suited to this sensitivity analysis application.

3.5.1 LIMITATIONS OF THE STUDY

The energy budget presented in this chapter does not represent all of the farm energy provided by the FEUS. This was because only those farming systems that are extensive users of farmland are relevant to the regional focus of CEEMA. The excluded energy consumers, including the horticultural enterprises such as market gardeners, fruit growers and greenhouses, are typically clustered within a few highly favourable climate zones, usually in proximity to population centres. In addition, relative to total agricultural energy use, these enterprises are very small and, consequently, small users of energy. In spite of the CEEMA data being derived from economic analysis, while the previous farm energy budget described by Dyer and Desjardins (2009) used actual crop statistics as input data, there was close agreement between these two sets of energy use estimates.

It should be cautioned that the farm energy budget described in this chapter will undergo changes, particularly since it applied to 2001. There are both uncertainties and on-going trends in several of the energy terms in this budget. The most dramatic case has been the impact of reduced tillage on farm use of diesel fuel for field operations (Dyer and Desjardins, 2005). An increasing popularity of diesel fuel for farm owned transport vehicles may mean that some use of diesel fuel for tasks other than field operations may have to be monitored and taken into account in future farm energy budget estimates. The fossil CO_2 emissions that can be attributed to farm use of electric power could also change as coal generating plants are replaced by natural gas, nuclear reactors, or renewable power sources. For example, natural gas, with its lower CO_2 to energy ratio than coal, is becoming increasingly available for this purpose (NEB, 2006).

There are also suggestions that ammonia-based nitrogen fertilizers could consume less natural gas than other forms of this chemical input (CAP, 2008) and that allowance for increased use of ammonia-based nitrogen fertilizer is needed in the carbon footprint of farm operations. However, the estimates of CO_2 emissions associated with the supply of farm chemical inputs by Dyer and Desjardins (2009), upon which this chapter was based, is consistent with, if not lower than, other studies. For example, over the four census years prior to 2001, the average national CO_2 emissions for chemical inputs reported by Dyer and Desjardins (2009) was 9% below the same period average fossil CO_2 emissions for this term by Janzen et al. (1999). Snyder et al. (2007) reported CO_2 to N conversion rates that were the same as the 4.05 t{CO_2}/t{N} conversion used by Dyer and Desjardins (2009) for Nebraska and 10% higher for Michigan.

The assumption that most farm animals are located near their feed sources was essential to the disaggregation of the three empirical energy terms. This assumption was sound for cattle as roughage makes up an important part of their diet and, except for drought years, its long-distance transport is uneconomic. This assumption was somewhat less sound for pork and poultry as feed grains (including oilseed meal) are more easily transported. Nevertheless, for these livestock types there is an advantage to having production near the cropland that provides the feed and is available for manure disposal. The higher spatial variation for the pork and poultry compared to beef and dairy in Table 3 would support the impact of this advantage. Although pork and poultry were the smallest of the five farming systems, and the three empirical energy terms were also the smallest terms, it would be worthwhile to gather data on the distances over which livestock farmers can cost-effectively ship feed grains. Furthermore, a reliable estimate of the energy used by farmers for transport would be essential to an objective carbon footprint comparison of livestock farming with biofuel feedstock production.

3.5.2 GOING FORWARD: IMPLICATIONS FOR BIOFUELS

Trends in farm energy levels will also reflect shifts in land use towards feedstock for biofuels. Providing farm type-specific energy data in Appendix A with this chapter identified the energy quantities that are most likely to shift as land resources are reallocated from livestock or food crops into feedstock if the biofuel market opportunities expand. Because of the uncertainties in the farm energy budget, such as more efficient manufacturing of farm inputs, and the land use challenges associated with the emerging biofuel industries, flexibility will be needed. The examples provided here with Figure 2 demonstrated how changes in land use can affect the area based intensity of farm energy consumption and fossil CO_2 emissions. Hence, the analytical procedures for farm energy described in this chapter are being maintained in a dynamic, integrated and repeatable computation procedure. With this flexibility it can facilitate revisions in the Canadian farm energy budget or shifts in farm management as predicted in an updated version of the CEEMA.

This chapter devoted relatively more space and effort to the electric power, gasoline and heating fuel terms than to the field work and two supply terms. Although the three terms from the FEUS were smaller energy quantities, there were two reasons for this extra attention. First, they have received almost no analysis, at least from a modeling perspective, prior this analysis. Consequently, the disaggregation of these terms was much more interpolative than process based. Second, the different levels of use by the five major farm types in Canada of these energy sources, combined with the regional differences in where these farming systems are most often found, resulted in the appreciable spatial variations at the CAR scale shown in Table 3, at least compared to Table 4.

The liquid fossil fuels burned in farm-owned vehicles (both gasoline and diesel) warrants more rigorous treatment because of its overlap with the question of the energy costs of transporting food products to processors and consumers, or feedstock to biofuel plants. Development of a predictive model for this term will depend on better understanding of how and where producers market their produce and the extent to which processers are involved in the collection of that produce, whether it is milk, wheat or canola oil. This is particularly true for biofuel feedstock where the haulage cost can grow in comparison to the production cost if the processing plants are not strategically located. Optimizing the locations of biofuel processing sites will depend on the knowledge of both energy uses and the spatial distribution of land use systems.

Much of the farm energy budget presented in this chapter was based on the 1996 FEUS. Including verification of the F4E2 model, five of the six terms in this energy budget were derived from this database. Updating the FEUS would also facilitate disaggregation of the later years in the farm fossil CO2 emissions budget described by Dyer and Desjardins (2009) to both the provinces and the CARS. The importance of farm energy in the GHG emissions budget for both agriculture and biofuels requires a repeat of the 1996 FEUS. Since the FEUS entailed survey methodology, rather than actual measurements, an updated FEUS would be an expensive undertaking in Canada. Whereas electric power showed some promise for a predictive tool (Dyer and Desjardins, 2006b), the other FEUS-based terms, gasoline and heating fuels, offer little hope of being worked into a predictive model, although they could be indexed to changing livestock populations. Fortunately, all three of these terms contribute relatively little to Canada's farm energy budget compared to the other three terms.

Growth in biofuel industries is driving the crop selections by many Canadian farmers towards feedstock crops. But as global population expands, major land use shifts will also occur in the food industries, such as from beef or pork production, to more grains and pulses for direct human consumption. Food industries that are now minor, such as vegetable production, may see dramatic growth in response to both food demand and to a warmer climate. Canadian agriculture may well be challenged by shortages of fossil fuel to do field work and commercial fertilizer. The CEEMA database also needs to be updated to help meet these challenges. Until a repetition of the FEUS is undertaken, updated regional farm energy use, and fossil CO_2 emission estimates using more recent census years and an up to date version of CEEMA, will help to fill the information gaps caused by looming changes in the sector.

REFERENCES

1. AAFC, 2011. Soil Landscapes of Canada (SLC), General Overview. Agriculture and Agri-food Canada (AAFC). http://sis.agr.gc.ca/cansis/nsdb/slc/intro.html. Accessed 7 May 2012.
2. Dyer, J.A. and R.L. Desjardins. 2003a. The impact of farm machinery management on the greenhouse gas emissions from Canadian agriculture. Journal of Sustainable Agriculture. 20(3):59-74.

3.   Dyer, J.A. and R.L. Desjardins. 2003b. Simulated farm fieldwork, energy consumption and related greenhouse gas emissions in Canada. Biosystems Engineering 85(4):503-513.

4.   Dyer, J.A. and R.L. Desjardins. 2005. Analysis of trends in CO2 emissions from fossil fuel use for farm fieldwork related to harvesting annual crops and hay, changing tillage practices and reduced summerfallow in Canada. Journal of Sustainable Agriculture. 25(3):141-156.

5.   Dyer, J.A. and R.L. Desjardins. 2005. Analysis of trends in CO2 emissions from fossil fuel use for farm fieldwork related to harvesting annual crops and hay, changing tillage practices and reduced summerfallow in Canada. Journal of Sustainable Agriculture. 25(3):141-156.

6.   Dyer, J.A. and R.L. Desjardins. 2006a. Carbon dioxide emissions associated with the manufacturing of tractors and farm machinery in Canada. Biosystems Engineering. 93(1):107-118.

7.   Dyer, J.A. and R.L. Desjardins, 2006b. An integrated index for electrical energy use in Canadian agriculture with implications for greenhouse gas emissions. Biosystems Engineering. 95(3):449-460.

8.   Dyer,J. A. and R. L. Desjardins. 2007. Energy-based GHG emissions from Canadian Agriculture. Journal of the Energy Institute. 80(2):93-95.

9.   Dyer, J.A. and R.L. Desjardins. 2009. A review and evaluation of fossil energy and carbon dioxide emissions in Canadian agriculture. Journal of Sustainable Agriculture 33(2):210-228.

10. Dyer, J.A., X.P.C. Vergé, R.L. Desjardins, D.E. Worth, B.G. McConkey. 2010a. The impact of increased biodiesel production on the greenhouse gas emissions from field crops in Canada. Energy for Sustainable Development 14(2):73–82.

11. Dyer, J.A., S.N. Kulshreshtha, B.G. McConkey and R.L. Desjardins. 2010b. An assessment of fossil fuel energy use and CO2 emissions from farm field operations using a regional level crop and land use database for Canada. Energy 35(5):2261-2269.

12. Dyer, J.A. O.Q. Hendrickson, R.L. Desjardins and H.L. Andrachuk. 2011a. An Environmental Impact Assessment of Biofuel Feedstock Production on Agro-Ecosystem Biodiversity in Canada. In: Agricultural Policies: New Developments. Chapter 3:87-115. Editor: Laura M. Contreras, ISBN 978-1-61209-630-8. Nova Science Publishers Inc. Hauppauge, NY 11788. 281 pp.

13. Dyer, J.A., X.P.C. Vergé, S.N. Kulshreshtha, R.L. Desjardins, and B.G. McConkey. 2011b. Residual crop areas and greenhouse gas emissions from feed and fodder crops that were not used in Canadian livestock production in 2001. Journal of Sustainable Agriculture 35(7):780-803.

14. Dyer, J.A., X.P.C. Vergé, R.L. Desjardins and B.G. McConkey. 2011c. Implications of biofuel feedstock crops for the livestock feed industry in Canada. In: Environmental Impact of Biofuels. Chapter 9, pages 161-178, Editor: Marco Aurelio Dos Santos Bernardes. InTech Open Access Publisher. Rijeka, Croatia. September 2011, ISBN 978-953-307-479-5. 270 pp.

15. CAP, 2008. Benchmark energy efficiency and carbon dioxide emissions. Canadian Ammonia Producers (CAP).Office of Energy Efficiency, Natural Resources Canada. Ottawa, Canada. ISBN 978-0-662-46150-0, Cat. No. M144-155/2007E-PDF. 34pp.

16. GAO, 2009. Biofuels – Potential effects and challenges of required increases in production and use. Report to Congressional Requesters. United States Government Accountability Office (GAO). August 2009. GAO-09-446. 184 pp.

17. Horner, G.L., J. Gorman, R.E. Howett, C.A. Carter and R.J. MacGregor. 1992. The Canadian regional agricultural model: structure, operation and development. Policy Branch. Technical Report 1/92. Ottawa. Agriculture and Agri-food Canada. http:www.agr.gc.ca/pol/index_e.php?s1=pub&s2=visions&page=v_pt2.

18. IFAS, 1998. Replacing hay with grain. Chapter 6, The Disaster Handbook - 1998 Edition. Institute of Food and Agricultural Sciences, Cooperative Extension Service, University of Florida.

19. Janzen, H.H., R.L. Desjardins, J.M.R. Asselin and B. Grace (Editors). 1999. The Health of our Air - Towards Sustainable Agriculture in Canada. Agriculture and Agri-food Canada. Publ. No. 1981 E. ISBN 0-662-27170-X, 105 pp.

20. Klein, K.K., LeRoy, D.G. 2007. The Biofuels Frenzy: What's in it for Canadian Agriculture? Green Paper Prepared for the Alberta Institute of Agrologists. Presented at the Annual Conference of Alberta Institute of Agrologists. Banff, Alberta, March 28, 2007. Department of Economics, University of Lethbridge. 46 pp.

21. Kulshreshtha, S.N. .B. Jenkins, R.L. Desjardins and J.C. Giraldez. 2000. A systems approach to estimation of greenhouse gas emissions from the Canadian agriculture and agri-food sector. World Resources Review. 12:321-337.

22. Malcolm, S. and M. Aillery. 2009. Growing crops for biofuels has spillover effects. Amber Waves 7(1):10-15. March issue.

23. Miller, J. I. and F. B. Morrison, 1950. Comparative Value of Mixed Hay and Corn Silage for Wintering Beef Cows. Journal of Animal Science. 9(2):243-247. http://jas.fass.org/content/9/2/243.abstract

24. Nagy, C.N. 2000. Energy and greenhouse gas coefficients inputs used in agriculture. Report to the Prairie Adaptation Research Collaborative (PARC). The Canadian Agricultural Energy End-use Data and Analysis Centre (CAEEDAC) and the Centre for Studies in Agriculture, Law and the Environment (CSALE).

25. NEB, 2006. Natural gas for power generation: issues and implications. ISBN 0-662-43472-2. Cat No. NE23-136/2006E. National Energy Board (NEB), Calgary, Alberta, Canada. 60pp. www.neb-one.gc.ca.

26. Neitzert, F., K. Olsen and P. Collas, 1999. Canada's greenhouse gas inventory - 1997- Emissions and removals with trends. Air Pollution Prevention Directorate, Environment Canada. Ottawa, Canada. ISBN 0-622-27783-X. Cat No. En49-8/5~9E. 160pp.

27. Neel, James B. 2012. Rules of thumb for winter feeding. Animal Science Info Series: AS-B-260, University of Tennessee, Agricultural Extension Service. http://animalscience.ag.utk.edu/beef/pdf/RulesOfThumb.260.pdf. Accessed 18 June 2012.

28. NRCan, 2005. Energy Efficiency Trends in Canada, 1990 and 1996 - 2003. Office of Energy Efficiency, Natural Resources Canada. Cat. No. M141-/2003. ISBN 0-662-68797-3, 64 pp.

29. Schoenian, Susan, 2011. The truth about grain: Feeding grain to small ruminants. Small Ruminant Info Sheet, University of Maryland Extension. http://www.sheepandgoat.com/articles/graintruth.html. Accessed 6 February 2012. Accessed 18 June 2012.

30. Snyder, C.S., T.W. Bruulsema and T.L. Jenzen, 2007. Greenhouse gas emission from cropping systems and the influence of fertilizer management. International Plant Research Institute, Norcross, Georgia, USA. 36 pp.
31. Statistics Canada, 2007, Census Agricultural Regions Boundary Files for the 2006 Census of Agriculture - Reference Guide. Catalogue no. 92-174-GIE. http://publications.gc.ca/collections/collection_2007/statcan/92-174-G/92-174-GIE2007000.pdf. Accessed 7 May 2012.
32. Tremblay, V. 2000. The 1997 Farm Energy Use Survey - Statistical Results. Contract Report prepared for The Office of Energy Use Efficiency, Natural Resources Canada, Ottawa, Ontario, Canada, and The Farm Financial Programs Branch, Agriculture and Agri-Food Canada, Ottawa, Ontario, Canada.
33. Vergé, X.P.C., Dyer, J.A., Desjardins, R.L. and Worth, D. 2007. Greenhouse gas emissions from the Canadian dairy industry during 2001. Agricultural Systems 94(3):683-693.
34. Vergé, X.P.C., Dyer, J.A., Desjardins, R.L. and Worth, D. 2008. Greenhouse gas emissions from the Canadian beef industry. Agricultural Systems 98(2):126–134.
35. Vergé, X.P.C., Dyer, J.A., Desjardins, R.L. and Worth, D. 2009a. Greenhouse gas emissions from the Canadian pork industry. Livestock Science. 121:92-101.
36. Vergé, X.P.C., Dyer, J.A., Desjardins, R.L. and Worth, D. 2009b. Long Term trends in greenhouse gas emissions from the Canadian poultry industry. Journal of Applied Poultry Research. 18: 210–222.
37. Vergé X.P.C., D.E. Worth, J.A. Dyer, R.L. Desjardins, and B.G. McConkey. 2011. LCA of animal production. In Green Technologies in Food Production and Processing, ed. by Joyce I. Boye and Yves Arcand, 83–114. New York: Springer.
38. Vergé, X.P.C., J.A. Dyer, D.E. Worth, W.N. Smith, R.L. Desjardins, and B.G. McConkey, 2012. A greenhouse gas and soil carbon model for estimating the carbon footprint of livestock production in Canada. Accepted in Animals.2:437-454

There are several supplemental files that are not available in this version of the article. To view this additional information, please use the citation on the first page of this chapter.

CHAPTER 4

Evaluating the Marginal Land Resources Suitable for Developing Bioenergy in Asia

JINGYING FU, DONG JIANG, YAOHUAN HUANG, DAFANG ZHUANG, AND WEI JI

4.1 INTRODUCTION

The world is facing problems related to finite availability of fossil fuels, the high price of petroleum, and the environmental impacts caused by the use of traditional fuels. The energy consumption of the world increased from 77,245 thousand barrels per day in 2001 to 88,034 thousand barrels per day in 2011. Asia Pacific accounted for 32% of the total world energy consumption [1]. This increase in energy demand is depleting fossil energy reserves at a high rate. In addition, the use of fossil fuels has caused many environmental problems, such as greenhouse gas (GHG) emissions. Therefore, energy security and climate change mitigation are two main drivers that have pushed renewable energy production to the top of the global agenda [2].

Evaluating the Marginal Land Resources Suitable for Developing Bioenergy in Asia. © Fu J, Jiang D, Huang Y, Zhuang D, and Ji W. Advances in Meterology **2014** (2014), http://dx.doi. org/10.1155/2014/238945. Licensed under Creative Commons Attribution 3.0 Unported License, http://creativecommons.org/licenses/by/3.0/.

Bioenergy, the most abundant and versatile type of renewable energy, has recently attracted worldwide attention [3]. Biofuels are environmentally friendly and carbon neutral and can play a prominent role in the energy portfolio [4]. The production of liquid biofuels can reduce GHG emissions by 12%–115% compared to traditional fossil fuels. GHG emissions are reduced 12% by the production and combustion of ethanol and 41% by biodiesel according to Hill et al. [5]. Adler et al. found that ethanol and biodiesel reduced GHG emissions by approximately 40% when derived from corn, by approximately 85% when from reed canary grass, and by approximately 115% when from hybrid switch grass and poplar [6]. The global warming potential (GWP, in kg CO_2-equivalent) of the production of biodiesel in the UK was calculated Stephenson et al. The results showed that large-scale production of biodiesel saved 26% of the GWP and small-scale production saved 32% of the GWP when compared to ultralow sulphur diesel [7].

The present global biomass demand for energy purposes is estimated to be 53 Quintillion joules [8]. Overall, global energy demand will grow 35%, even with significant efficiency gains. Energy demand in developing nations will rise 65 percent by 2040 (compared to 2010) as a result of expanding economies and growing populations. According to the new public energy outlook, 75 percent of the world's population will reside in Asia Pacific and Africa by 2040. India will have the largest population after 2030 [9]. A wide range of indicators suggest that dramatic developments are taking place in Asian energy markets [10], and large-scale bioenergy development is extremely urgent.

Recently, a number of studies have assessed the potential of biofuel. Kumar et al. assessed ethanol and biodiesel development in Thailand in terms of feedstock, production, planned targets, policies, and sustainability (environmental, socioeconomic, and food security aspects) [11]. An assessment of bioenergy potential was also carried out in England, the Midwest United States, China, and other countries [12–15]. The environmental life cycle assessment of lignocellulosic conversion to ethanol was reviewed by Borrion et al. Numerous studies of lignocellulosic ethanol fuel generated significantly different results due to differences in data, methodologies, and local geographic conditions [16]. In addition to feedstock, energy benefits, and GHG reductions, issues related to land resources and food security are an important consideration for Asia-scaled applications.

Schröder et al. considered bioenergy development as an effective way to save the world from an energy crisis. They illustrated the ability to produce novel energy plants for growth on abandoned land [17]. Liu et al. analyzed the bioenergy production potential on marginal land in Canada. The results showed that approximately 9.48 million hectares could be identified as available marginal land in Canada. If this land was fully utilized for growing energy crops, the production of biofuel would be 33 million tons (using switch grass) or 380 million tons (using hybrid poplar). Batidzirai et al. reviewed the current, state-of-the-art approaches and methodologies used in bioenergy assessments and identified key elements that are critical determinants of bioenergy potentials. Bioenergy potential assessments in the US, China, India, Indonesia, and Mozambique were also presented in the paper [18]. Hattori and Morita [19] studied which energy crops can be used for sustainable bioethanol production and where they can be grown. They found that, in Japan and other Asian countries, rice can be grown as an energy crop in unused low-land paddy fields. Bioenergy development in China has also been studied, especially the potential energy production on marginal land in the context of food security [3, 20–22]. The biomass plant *Jatropha curcas L.* (*JCL*) was shown to be a better economic, environmental, and land preservation alternative to corn or millet planted in the poor, gravel soil. and drought land in Taitung, Taiwan [23].

However, the bioenergy development in the above-mentioned studies and most other current research is studied on a regional scale. A potential bioenergy view of the entirety of Asia is not available. The main objective of this study is to present a comprehensive assessment on the marginal land resources which are suitable for developing bioenergy in Asia, without affecting food and ecoenvironmental security. Asia is the world's largest and most populous continent. It is facing significant pressure for food production. To avoid using the limited amount of arable land, adaptable energy plants that can be grown on marginal land and at scale must be used. *Cassava*, *P. chinensis*, and *JCL* have been widely proven in existing literature and are further studied in this paper [14, 24–34]. *P. chinensis* and

JCL are nonfood plants. *Cassava* is used as a food plant in some places. However, we only analyzed its development potential in uncultivated areas (marginal land).

Cassava and *JCL* are classified as second-generation biofuel feedstock, which are derived from crop residues, energy plants, and construction waste [35]. They can reduce GHG emissions and energy dependency during the life cycle when compared to the fossil fuel. The most important advantage of second-generation biofuels is that they will ensure the security of food supply compared with first-generation biofuels which are produced from food-based crops. They are sustainable and environmentally friendly [36]. Bioethanol is produced by hydrolysis and fermentation of carbohydrate feedstock. This type of energy plant usually has high saccharide, starch, and fiber content. *Cassava* which has been widely studied is this kind of plant. Biodiesel is produced from oil plants such as *JCL*. The oil extracted is blended with diesel to produce fuel [34, 37–47].

To achieve our goal, we used Geographic Information System (GIS) technology to identify the spatial distribution of marginal lands which are suitable for bioenergy development. The datasets of growth habits of energy plants, remote sensing-derived land cover, terrain, meteorological data, and soil data were processed to 1 km^2 grid across Asia.

4.2 METHODOLOGY

Four steps were implemented for this study. First, we identified the marginal land resources suitable for developing bioenergy in Asia. Second, we chose the three aforementioned energy plants that have been proven as biofuels. Third, we reviewed the environmental requirements of each energy plant including preferred meteorological conditions, soil, and terrain. Finally, a multiple factor analysis method was used to evaluate the bioenergy development potential based on the availability of marginal land resources and the growing conditions of the energy plants within the data grid. This analysis was performed using ArcMap software. The specific procedures are presented in Figure 1.

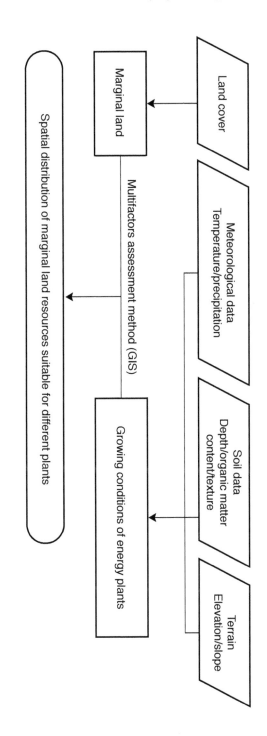

FIGURE 1: Evaluation of spatial distribution and suitability of marginal land resources for energy plants.

4.2.1 DATA ACQUISITION

In this study, the land cover, terrain (including elevation and slope), meteorological conditions (including precipitation and temperature), and soil data (including soil organic matter content, soil depth, and soil texture) were used. The data sources and spatial resolutions are listed in Table 1. All the data in this study was resampled to cover the entirety of Asia at a 1 km² resolution.

TABLE 1: Input data for identification of marginal land resources.

Input data	Data sources	Original spatial resolution
Land cover	ESA 2010 and UCLouvain [48]	1 km
Terrain		
Elevation	SRTM [49]	90 m
Slope		90 m
Meteorological data		
Precipitation	WorldClim [50]	30 arc-seconds (~1 km)
Temperature		30 arc-seconds (~1 km)
Soil data		
Organic matter content	FAO/IIASA/ISRIC/ISS-CAS/JRC [51]	30 arc-seconds (~1 km)
Soil depth		30 arc-seconds (~1 km)
Soil texture		30 arc-seconds (~1 km)

4.2.1.1 LAND COVER

Land resources defined as marginal must also include land that is considered economically marginal. Therefore, we spatially define marginal land resources based on the land cover classification of unused land. The land cover dataset can be obtained from the GlobCover project. There are 23 land cover types in the dataset. This is the fundamental dataset for identification of marginal land that is suitable for bioenergy development.

4.2.1.2 TERRAIN

The CGIAR-CSI GeoPortal provides SRTM 90 m digital elevation data for the entire world [49]. The digital elevation models (DEMs) of Asia were extracted from the dataset above, and the slope was calculated using the spatial analysis tool in ArcMap. Thresholds for DEMs and slope, based on the growth habits of each energy plant, were determined (see Section 2.2).

4.2.1.3 METEOROLOGICAL DATA

WorldClim is a set of global climate layers (climate grids) with a spatial resolution of 30 arc-seconds (often referred to as 1 km resolution). The precipitation and temperature data used in this study were interpolated from observed data from 1950 to 2000 [50]. These two elements are very important for identifying suitable land. The requirement of each energy plant was identified (see Section 2.3).

4.2.1.4 SOIL DATA

The Harmonized World Soil Database (HWSD) contributes sound scientific knowledge for planning sustainable expansion of agricultural production and for guiding policies to address emerging land competition issues concerning food production, bioenergy demand, and threats to biodiversity. A resolution of approximately 1 km was selected to analyze agroecological zoning, food security, and climate change impacts. Soil attribute data were linked with GIS so that specific parameters could be displayed, characterized, and analyzed. These parameters include soil units, organic carbon, pH, water storage capacity, soil depth, cation exchange, clay fraction, total exchangeable nutrients, lime and gypsum contents, sodium exchange percentage, salinity, textural class, and granulometry [51]. Soil texture, organic carbon content, and depth are key factors for growing energy plants.

4.2.2 IDENTIFICATION OF MARGINAL LAND

Marginal land has various meanings in different disciplines and, therefore, the spatial coverage of marginal land differs. Generally, marginal land is evaluated in terms of a cost-benefit analysis and is determined to be economically marginal [3]. Zhuang et al. established a marginal land evaluation system based on the definition of the Ministry of Agriculture (MoA) of China, a qualitative analysis of energy plants in different parts of China, expert suggestions on local planting of energy plants, land resources, and ecology, and other factors [22]. According to the definition of marginal land by MoA of China, marginal land is winter-fallowed paddy land and wasteland that may be used to cultivate energy crops. We only considered the wasteland in this study. Wasteland includes natural grassland, sparse forestland, scrubland, and unused land that may be used to grow energy crops [3]. We selected six land cover types as the available marginal land for growing the energy plants in compliance with the principle that bioenergy development should not compete with cropland and ecologically protected land. These six types were "mosaic vegetation (grassland/shrub land/forest) (50–70%)/cropland (20–50%)," "sparse (<15%) vegetation," "mosaic grassland (50–70%)/forest or shrub land (20–50%)," "closed to open (>15%) (broad-leaved or needle-leaved and evergreen or deciduous) shrub land (<5 m)," "closed to open (>15%) grassland or woody vegetation on regularly flooded or waterlogged soil," and "bare areas." The selection of land cover types for each country can be flexible based on the law, policy, environmental conditions, and special regulations. For example, nature reserves should be excluded in further studies.

4.2.3 CHARACTERISTICS OF SELECTED ENERGY PLANTS

Cassava, as feedstock for fuel ethanol, has three advantages over others. First, *Cassava* is a shrubby tropical plant, widely grown for its large, tuberous, starchy roots, especially on marginal land. Second, *Cassava* is not a staple food for most people in Asia. Third, it is easy to comminute, has short cooking times, and has a low gelatinization temperature. Therefore, *Cassava* is a suitable feedstock for fuel ethanol [33, 57].

P. chinensis is an ideal species for producing biodiesel. The tree has several outstanding characteristics: drought resistance, tolerance to cold climate, and tolerance to poor, acid, or alkaline soils. It also has some advantages that cannot be replaced by other trees, such as oil yield and its conversion rate, biodiesel quality, geographical distribution, adaptability, and economic benefits cycle. Therefore, *P. chinensis* is considered an important source of biodiesel [34, 47, 52].

TABLE 2: Growing conditions of energy plants.

Growing conditions	Cassava [24]		P. chinensis [52, 53]		JCL [54–56]	
	Suitable	Moderately suitable	Suitable	Moderately suitable	Suitable	Moderately suitable
Meteorological data						
Annual average temperature/°C	21~29	18~21	10~15.3	5.8~10 or 15.3~28.4	20~25	17~20
Average annual extreme lowest temperature/°C	—	—	≥−15	−26.5~−15	≥2	0~2
Accumulated temperature of 10/°C·d	—	—	≥3800	1180~3800	—	—
Precipitation/mm	1000~2000	600~1000 or 2000~6000	400~1300	1300~1900	600~1000	300~600 or 1000~1300
Soil data [51]						
Soil depth/cm	≥75	30~75	≥60	30~60	≥75	30~75
Soil organic matter content/%	≥3.5	1.5~3.5	—	—	≥3.5	1.5~3.5
Soil texture/classes	1	2	—	—	1	2
Terrain						
Elevation/m	≤1500	1500~2000	—	—	≤500	500~1600
Slope/°	≤15	15~25	≤15	15~25	≤15	15~25

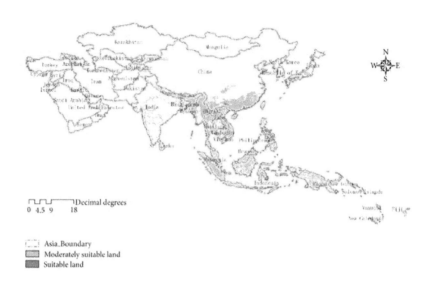

FIGURE 2: Distribution of marginal land resources for *Cassava*.

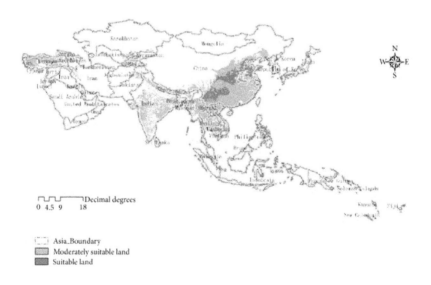

FIGURE 3: Distribution of marginal land resources for *P. chinensis.*

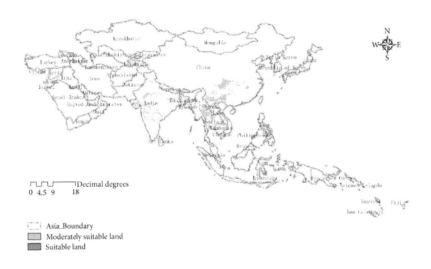

FIGURE 4: Distribution of marginal land resources for JCL.

JCL is a famous biofuel plant and has been studied globally [31, 58]. It is a tropical species, native to Mexico and Central America, but is widely distributed in wild or semicultivated stands in Latin America, Africa, India, and South-East Asia [59]. The *Jatropha curcas* plant is a nonedible, drought-resistant, perennial plant that has the capability to grow on marginal lands because it requires very few nutrients to survive [34, 44]. *Jatropha* has several other advantages, such as a short gestation period, resistance to common pests, lack of consumption by cattle, and production of biofertilizer and glycerine as useful by-products of biodiesel. In addition, the seed collection period of *Jatropha* does not coincide with the rainy season in June and July, which is when most agricultural activities take place. This makes it possible for people to generate additional income during the slack agricultural season [60, 61].

All the specific requirements of the energy plants were chosen according to the literature and advice from experts. The growing conditions of the energy plants are listed in Table 2. The marginal land presented in the previous section was used as the basic condition in the multiple factor analysis.

We used strict criteria during the identification of suitable and moderately suitable areas for energy plants. Marginal land resource areas were only characterized as suitable if all of the suitable conditions were met. If one of the growing conditions was moderately suitable, the land resources were identified as moderately suitable.

The soil texture data used in this paper was classified into two classes. Class 1 was defined as fine textured with more than 35% clay. Class 2 was defined as medium textured with a clay percentage between 18% and 35%. The soil texture requirement of energy plants is that the volumetric ratio of clay should be more than 30% for suitable land and 18%~35% for moderately suitable land. Therefore, there may be more potential land resource areas available for growing energy plants if more accurate soil data can be obtained.

4.3 RESULTS AND DISCUSSION

The planting zones of each energy plant were identified based on marginal land areas and the plant's growth habits. The multiple factor analysis method was adopted to evaluate the suitable marginal land resources based on the evaluation criteria for suitable and moderately suitable growing conditions of each single factor and the type of available land cover. The distributions of marginal land resources suitable for the three energy plants are presented in Figures 2, 3, and 4.

From Figures 2–4, we can see that the area suitable for the growth of *P. chinensis* is much larger than those for the other two plants. Approximately 70% of Asian countries have more than one thousand square kilometers of marginal land resources suitable for *P. chinensis*. *Cassava* and *JCL* resources are limited because they require warmer temperatures than *P. chinensis*, and *Cassava* has a higher precipitation requirement. The results in Table 3 show that the areas of marginal land resources of *Cassava*, *P. chinensis*, and *JCL* are nearly 1.12 million, 2.41 million, and 237 thousand square kilometers, respectively. China has the most marginal land area available for all of the energy plants. Myanmar possesses 21% of the land resources suitable for *Cassava*. Turkey and Thailand have the second largest marginal land resources suitable for *P. chinensis* and *JCL*. Shrub land is the dominant land cover type for growing energy plants, which accounts for 51.14% of the total suitable area. Mosaic vegetation is next, accounting for 34.49%.

TABLE 3: Marginal land resources suitable and moderately suitable for *Cassava*, *P. chinensis*, and *JCL* planting based on multiple factor analysis in Asia (km²).

Land cover	Cassava		P. chinensis		JCL		Total	
	S	M	S	M	S	M	S	M
Mosaic vegetation	1422	307537	130443	769321	1	92458	131866	1169316
Mosaic grassland	4	3697	17223	73886	0	2461	17227	80044
Shrub land	2089	788006	73008	942428	2	123672	75099	1854106
Herbaceous vegetation	6	16928	9221	89459	0	8794	9227	115181
Sparse vegetation	0	684	28002	180892	0	5732	28002	187308
Bare areas	0	3572	7529	88703	0	4530	7529	96805
Total	3521	1120424	265426	2144689	3	237647	268950	3502760

S: suitable land; M: moderately suitable land.

4.4 CONCLUSION

In this paper, a multiple factor analysis method was adopted to identify marginal land resources for three types of energy plants (*Cassava*, P. chinensis, and *JCL*) in Asia based on land cover, meteorological data, soil characteristics, terrain data, and the growth habits of energy plants. GIS was used to identify potential land resource areas at the resolution of 1 square kilometer. The conclusions of this study are as follows.

1. The areas of marginal land suitable for *Cassava*, *P. chinensis*, and *JCL* were established to be 1.12 million, 2.41 million, and 0.237 million km², respectively. The policy and environmental constraints of each specific county were not considered in this study.

2. China has great prospects for bioenergy development. It has the most marginal land resources available for all three energy plants. Myanmar, Turkey, and Thailand have the second largest areas of marginal land resources available for *Cassava*, *P. chinensis*, and *JCL*, respectively.

3. With regard to land cover, shrub land is the dominant land cover type for growing energy plants, accounting for 51.14% of the total suitable area. Mosaic vegetation is second, accounting for 34.49%.

Bioenergy development is important and full of challenges. Further research needs to be performed to choose the best feedstock, improve marginal land resource calculations using more accurate input data, estimate the energy production potential, and analyze the environmental effects coupled with social and economic benefits.

REFERENCES

4. BP Statistical Review of World Energy, 2012.
5. A. Karp and I. Shield, "Bioenergy from plants and the sustainable yield challenge," New Phytologist, vol. 179, no. 1, pp. 15–32, 2008.
6. Y. Tang, J.-S. Xie, and S. Geng, "Marginal land-based biomass energy production in China," Journal of Integrative Plant Biology, vol. 52, no. 1, pp. 112–121, 2010.
7. J. Zhuang, R. W. Gentry, G.-R. Yu, G. S. Sayler, and J. W. Bickham, "Bioenergy sustainability in China: potential and impacts," Environmental management, vol. 46, no. 4, pp. 525–530, 2010.
8. J. Hill, E. Nelson, D. Tilman, S. Polasky, and D. Tiffany, "Environmental, economic, and energetic costs and benefits of biodiesel and ethanol biofuels," Proceedings of the National Academy of Sciences of the United States of America, vol. 103, no. 30, pp. 11206–11210, 2006.
9. P. R. Adler, S. J. del Grosso, and W. J. Parton, "Life-cycle assessment of net greenhouse-gas flux for bioenergy cropping systems," Ecological Applications, vol. 17, no. 3, pp. 675–691, 2007.
10. A. L. Stephenson, J. S. Dennis, and S. A. Scott, "Improving the sustainability of the production of biodiesel from oilseed rape in the UK," Process Safety and Environmental Protection, vol. 86, no. 6, pp. 427–440, 2008.
11. REN21 Renewables 2012 Global Status Report. Renewable Energy Policy Network for the 21st Century, 2012.
12. ExxonMobil, The Outlook for Energy: A View to 2040, 2013.
13. M. M. Mochizuki, A. Tellis, and M. Wills, Confronting Terrorism in the Pursuit of Power: Strategic Asia, 2004-2005, 2004.

14. S. Kumar, P. Abdul Salam, P. Shrestha, and E. K. Ackom, "An assessment of Thailand's biofuel development," Sustainability, vol. 5, no. 4, pp. 1577–1597, 2013.

15. A. Thomas, A. Bond, and K. Hiscock, "A GIS based assessment of bioenergy potential in England within existing energy systems," Biomass & Bioenergy, vol. 55, pp. 107–121, 2013.

16. I. Gelfand, R. Sahajpal, X. Zhang, R. C. Izaurralde, K. L. Gross, and G. P. Robertson, "Sustainable bioenergy production from marginal lands in the US Midwest," Nature, vol. 493, no. 7433, pp. 514–517, 2013.

17. S. Liang, M. Xu, and T. Z. Zhang, "Life cycle assessment of biodiesel production in China," Bioresource Technology, vol. 129, pp. 72–77, 2013.

18. Y. P. Wu, S. G. Liu, and Z. P. Li, "Identifying potential areas for biofuel production and evaluating the environmental effects: a case study of the James River Basin in the Midwestern United States," Global Change Biology Bioenergy, vol. 4, no. 6, pp. 875–888, 2012.

19. A. L. Borrion, M. C. McManus, and G. P. Hammond, "Environmental life cycle assessment of lignocellulosic conversion to ethanol: a review," Renewable & Sustainable Energy Reviews, vol. 16, no. 7, pp. 4638–4650, 2012.

20. P. Schröder, R. Herzig, B. Bojinov et al., "Bioenergy to save the world: producing novel energy plants for growth on abandoned land," Environmental Science and Pollution Research, vol. 15, no. 3, pp. 196–204, 2008.

21. B. Batidzirai, E. M. W. Smeets, and A. P. C. Faaij, "Harmonising bioenergy resource potentials-methodological lessons from review of state of the art bioenergy potential assessments," Renewable & Sustainable Energy Reviews, vol. 16, no. 9, pp. 6598–6630, 2012.

22. T. Hattori and S. Morita, "Energy crops for sustainable bioethanol production; which, where and how?" Plant Production Science, vol. 13, no. 3, pp. 221–234, 2010.

23. Q. Zhang, J. Ma, G. Qiu et al., "Potential energy production from algae on marginal land in China," Bioresource Technology, vol. 109, pp. 252–260, 2012.

24. H. Qiu, J. Huang, M. Keyzer et al., "Biofuel development, food security and the use of marginal land in china," Journal of Environmental Quality, vol. 40, no. 4, pp. 1058–1067, 2011.

25. D. Zhuang, D. Jiang, L. Liu, and Y. Huang, "Assessment of bioenergy potential on marginal land in China," Renewable & Sustainable Energy Reviews, vol. 15, no. 2, pp. 1050–1056, 2011.

26. Y.-K. Tseng, "The economical and environmental advantages of growing Jatropha curcas on marginal land," Advanced Materials Research, vol. 361–363, pp. 1495–1498, 2012.

27. Z.-C. Li and X.-M. Liang, "Analysis of the potential of *cassava* used as raw materials for fuel alcohol production in China," Liquor-Making Science & Technology, vol. 4, pp. 31–33, 2010.

28. C. Jansson, A. Westerbergh, J. Zhang, X. Hu, and C. Sun, "*Cassava*, a potential biofuel crop in China," Applied Energy, vol. 86, no. 1, pp. S95–S99, 2009.

29. C. Sorapipatana and S. Yoosin, "Life cycle cost of ethanol production from *cassava* in Thailand," Renewable & Sustainable Energy Reviews, vol. 15, no. 2, pp. 1343–1349, 2011.

30. K. Sriroth, K. Piyachomkwan, S. Wanlapatit, and S. Nivitchanyong, "The promise of a technology revolution in *cassava* bioethanol: from Thai practice to the world practice," Fuel, vol. 89, no. 7, pp. 1333–1338, 2010.
31. S. Kumar, J. Singh, S. M. Nanoti, and M. O. Garg, "A comprehensive Life Cycle Assessment (LCA) of Jatropha biodiesel production in India," Bioresource Technology, vol. 110, pp. 723–729, 2012.
32. Y.-K. Tseng, "The economical and environmental advantages of growing Jatropha curcas on marginal land," Advanced Materials Research, vol. 361-363, pp. 1495–1498, 2012.
33. H. C. Ong, T. M. I. Mahlia, H. H. Masjuki, and R. S. Norhasyima, "Comparison of palm oil, Jatropha curcas and Calophyllum inophyllum for biodiesel: a review," Renewable & Sustainable Energy Reviews, vol. 15, no. 8, pp. 3501–3515, 2011.
34. K. Openshaw, "A review of Jatropha curcas: an oil plant of unfulfilled promise," Biomass and Bioenergy, vol. 19, no. 1, pp. 1–15, 2000.
35. L. Axelsson, M. Franzén, M. Ostwald, G. Berndes, G. Lakshmi, and N. H. Ravindranath, "Jatropha cultivation in southern India: assessing farmers' experiences," Biofuels, Bioproducts and Biorefining, vol. 6, no. 3, pp. 246–256, 2012.
36. D. Zhuang, D. Jiang, L. Liu, and Y. Huang, "Assessment of bioenergy potential on marginal land in China," Renewable & Sustainable Energy Reviews, vol. 15, no. 2, pp. 1050–1056, 2011.
37. M. Tang, P. Zhang, L. Zhang, M. Li, and L. Wu, "A potential bioenergy tree: Pistacia chinensis Bunge," in Proceedings of the International Conference on Future Energy, Environment, and Materials B, G. Yang, Ed., pp. 737–746, 2012.
38. B. Antizar-Ladislao and J. L. Turrion-Gomez, "Second-generation biofuels and local bioenergy systems," Biofuels, Bioproducts and Biorefining, vol. 2, no. 5, pp. 455–469, 2008.
39. A. K. Chandel, E. C. Chan, R. Rudravaram, M. L. Narasu, L. V. Rao, and P. Ravinda, "Economics and environmental impact of bioethanol production technologies: an appraisal," Biotechnology and Molecular Biology Review, vol. 2, no. 1, pp. 14–32, 2007.
40. H. Shao and L. Chu, "Resource evaluation of typical energy plants and possible functional zone planning in China," Biomass and Bioenergy, vol. 32, no. 4, pp. 283–288, 2008.
41. C. Zhang, W. Han, X. Jing, G. Pu, and C. Wang, "Life cycle economic analysis of fuel ethanol derived from *cassava* in southwest China," Renewable & Sustainable Energy Reviews, vol. 7, no. 4, pp. 353–366, 2003.
42. E. Nuwamanya, L. Chiwona-Karltun, R. S. Kawuki, and Y. Baguma, "Bio-ethanol production from non-food parts of *cassava* (Manihot esculenta Crantz)," Ambio, vol. 41, no. 3, pp. 262–270, 2012.
43. T. Silalertruksa and S. H. Gheewala, "Security of feedstocks supply for future bioethanol production in Thailand," Energy Policy, vol. 38, no. 11, pp. 7476–7486, 2010.
44. R. Sarin, M. Sharma, S. Sinharay, and R. K. Malhotra, "Jatropha-palm biodiesel blends: an optimum mix for Asia," Fuel, vol. 86, no. 10-11, pp. 1365–1371, 2007.
45. S. Jain and M. P. Sharma, "Prospects of biodiesel from Jatropha in India: a review," Renewable & Sustainable Energy Reviews, vol. 14, no. 2, pp. 763–771, 2010.

46. R. Abdulla, E. S. Chan, and P. Ravindra, "Biodiesel production from Jatropha curcas: a critical review," Critical Reviews in Biotechnology, vol. 31, no. 1, pp. 53–64, 2011.

47. C.-Y. Yang, Z. Fang, B. Li, and Y.-F. Long, "Review and prospects of Jatropha biodiesel industry in China," Renewable & Sustainable Energy Reviews, vol. 16, no. 4, pp. 2178–2190, 2012.

48. M. A. Kalam, J. U. Ahamed, and H. H. Masjuki, "Land availability of Jatropha production in Malaysia," Renewable & Sustainable Energy Reviews, vol. 16, no. 6, pp. 3999–4007, 2012.

49. M. H. Chakrabarti, M. Ali, J. N. Usmani, et al., "Status of biodiesel research and development in Pakistan," Renewable & Sustainable Energy Reviews, vol. 16, no. 7, pp. 4396–4405, 2012.

50. L. Lu, D. Jiang, D. Zhuang, and Y. Huang, "Evaluating the marginal land resources suitable for developing pistacia chinensis-based biodiesel in China," Energies, vol. 5, no. 7, pp. 2165–2177, 2012.

51. S. Bontemps, P. Defourny, E. van Bogaert, et al., GlobCover 2009, vol. 53, European Spatial Agency-Université Catholique de Louvain, 2011.

52. A. Jarvis, H. I. Reuter, A. Nelson, and E. Guevara, "Hole-filled SRTM for the globe version 4," The CGIAR-CSI SRTM 90m Database, 2008, http://www.cgiar-csi.org.

53. R. J. Hijmans, S. E. Cameron, J. L. Parra, P. G. Jones, and A. Jarvis, "Very high resolution interpolated climate surfaces for global land areas," International Journal of Climatology, vol. 25, no. 15, pp. 1965–1978, 2005.

54. FAO/IIASA/ISRIC/ISS-CAS/JRC Harmonized World Soil Database (Version 1.1), 2009.

55. X. Hou, H. Zuo, and H. Mou, "Geographical distribution of energy plant Pistacia chinensis Bunge in China," Ecology and Environmental Sciences, vol. 19, pp. 1160–1164, 2010.

56. Y. Fu, X. Pan, and H. Gao, "Geographical distribution and climate characteristics of habitat of Pistacia chinensis Bunge in China," Chinese Journal of Agrometeorology, vol. 30, no. 3, pp. 318–322, 2009.

57. J. Heller, Physic Nut, Jatropha curcas L., vol. 1, Bioversity International, 1996.

58. R. K. Henning, "Jatropha curcas L," Plant Resources of the Tropical Africa, vol. 14, pp. 116–122, 2004.

59. K. Eckart and P. Henshaw, "Jatropha curcas L. and multifunctional platforms for the development of rural sub-Saharan Africa," Energy for Sustainable Development, vol. 16, no. 3, pp. 303–311, 2012.

60. H. Yang, L. Chen, Z. Yan, and H. Wang, "Emergy analysis of cassava-based fuel ethanol in China," Biomass and Bioenergy, vol. 35, no. 1, pp. 581–589, 2011.

61. N. Foidl, G. Foidl, M. Sanchez, M. Mittelbach, and S. Hackel, "Jatropha curcas L. as a source for the production of biofuel in Nicaragua," Bioresource Technology, vol. 58, no. 1, pp. 77–82, 1996.

62. V. C. Pandey, K. Singh, J. S. Singh, A. Kumar, B. Singh, and R. P. Singh, "Jatropha curcas: a potential biofuel plant for sustainable environmental development," Renewable & Sustainable Energy Reviews, vol. 16, no. 5, pp. 2870–2883, 2012.

63. S. Kumar, A. Chaube, and S. K. Jain, "Sustainability issues for promotion of Jatropha biodiesel in Indian scenario: a review," Renewable & Sustainable Energy Reviews, vol. 16, no. 2, pp. 1089–1098, 2012.

64. P. Kumar Biswas, S. Pohit, and R. Kumar, "Biodiesel from jatropha: can India meet the 20% blending target?" Energy Policy, vol. 38, no. 3, pp. 1477–1484, 2010.

CHAPTER 5

Energy Potential of Biomass from Conservation Grasslands in Minnesota, USA

JACOB M. JUNGERS, JOSEPH E. FARGIONE, CRAIG C. SHEAFFER, DONALD L. WYSE, AND CLARENCE LEHMAN

5.1 INTRODUCTION

Perennial biomass is an alternative to conventional starch-based biofuel feedstocks such as corn. It may improve land-use efficiency, reduce greenhouse gas emissions, promote biodiversity, and support other components of sustainability [1]–[3]. Research comparing ecosystem services of various native and non-native perennial bioenergy crops in the Upper Midwest indicates that bioenergy systems with more plant species support greater avian diversity [4], abundance and diversity of beneficial arthropods [5], carbon storage and complexity of belowground food webs [6]. In many regions of North America, diverse grasslands have not produced as much gross biomass as dedicated energy crops grown in monoculture such as switchgrass [7]. This has initiated questions regarding the economic viability of diverse grassland bioenergy, yet few studies have quantified bioenergy yields from diverse perennial plantings over multiple years. Only recently have studies compared the bioenergy potential of mixed-species grasslands harvested with production-scale techniques in various regions of the Upper Midwest [8].

Growing biomass on land unsuitable for commodity crops transforms the economic outlook for bioenergy systems. Bioenergy production from feedstocks grown on marginal or underutilized land, such as land enrolled in the Conservation Reserve Program (CRP), can provide immediate greenhouse gas benefits [9] while avoiding competition for land between food and energy crops [10]. One idea is to harvest biomass from CRP land as revenue to supplement government subsidies, potentially incentivizing renewal of CRP contracts and offsetting recent trends in expiring CRP acreage [11]. Current CRP regulations do not allow biomass harvest from land enrolled in the program. If economic opportunities from bioenergy initiate new regulations that allow biomass harvest, these regulations should be designed to support the original intentions of the CRP, including improved wildlife abundance [12], an important component of biodiversity.

Other conservation lands managed for wildlife by state, federal, and non-profit agencies have been planted with mixtures of perennial grassland species. These may serve as biomass sources for energy production. Studies are underway to determine the effects of biomass harvest on resident wildlife in various types of conservation grasslands [13]. If research concludes that conservation grasslands can be managed for bioenergy and biodiversity simultaneously, then the quality and quantity of harvested biomass from conservation lands should be considered before bioenergy management is implemented.

The amount of bioenergy from conservation grasslands depends on both biomass quantity and quality. One means of measuring biomass quantity is to multiply yields from CRP fields in different regions of North America by estimates of available acreage [8], [14]–[16]. These yields can then be extrapolated to estimate biomass from land not currently enrolled in, but eligible for conservation programs. Another important component of predicting bioenergy potential is biomass quality, often defined by the mineral and sugar concentrations of the biomass. Mineral concentrations are used to predict conversion efficiency for thermochemical energy production. High concentrations of alkali metals in post-combustion ash lead to slagging and fouling in thermochemical systems [17], while high concentrations of N, S, and other elements pose issues of oxide emissions and possibly nutrient removal from soils in long-term harvested systems [18]. Predicting the efficiency of biofuel production with biochemical technologies requires measuring the plant sugar and carbohydrate concentrations. High values of cellulose and hemicellulose relative to lignin results in greater liquid biofuel potential [19].

Variation in the quantity and quality of grassland biomass with respect to energy production—hereafter called bioenergy potential—can occur due to variation in plant species composition, geographic location, and management activities. Plant composition influences bioenergy potential with studies indicating positive relationships between (i) biomass yield and planted species richness [2] and (ii) relative cover of warm-season grasses (C4) and lignocellulose ratios that favor ethanol production [14]. In southern Iowa, spatial variation in biomass yield and elemental composition was greater within fields than between fields and was correlated to individual species within cool-season (C3) grasslands [20]. A broad-scale analysis of switchgrass yields across the Great Plains indicated that within-field variation is small enough to consider the mean biomass yield of a field for modeling purposes [21]. Di Virgilio et al. found correlations between switchgrass yields and both soil fertility and moisture, which were interpreted as sources of within-field variation [22].

Management activities, including harvest, also affect bioenergy potential. Harvesting biomass after senescence allows for plants to translocate nutrients to belowground tissues, but harvesting post-senescence means that vegetation is removed after peak biomass and lodging have occurred. In Oklahoma and South Dakota, delaying harvest until October increased yields and decreased N and ash concentrations in CRP biomass compared to pre-peak biomass harvests [16], [23]. Harvesting switchgrass-dominated CRP lands every year compared with alternate years increased yields [24], while deferring harvest to more than two year intervals lowered bioenergy potential in Canadian conservation grasslands managed for wildlife [25].

In the present study, we modeled bioenergy potential of conservation grasslands based on three response variables related to quantity and quality: biomass yield, theoretical ethanol conversion efficiency, and plant tissue N. We used data collected from large-scale plots distributed across three locations of western Minnesota and harvested with commercial-scale tools and techniques. Our objectives were (i) to determine biomass yields, theoretical ethanol conversion efficiency, and plant tissue N content from conservation grasslands, (ii) to measure the variability of bioenergy potential along a latitudinal gradient in western Minnesota, and (iii) to understand what factors affect bioenergy potential by modeling the three response variables with data on plant communities, soil fertility, precipitation, and management activities while accounting for space and time. Two harvest treatments were used to determine if yields from completely harvested plots followed similar trends through time as yields from plots that included previously unharvested regions of biomass. Our results are intended to aid policy and land-management decisions regarding the use of conservation grasslands for bioenergy production in the Upper Midwest, USA.

5.2 METHODS

5.2.1 EXPERIMENTAL DESIGN

In 2008, we located and delineated 60 plots within existing grasslands enrolled in a conservation program. Plots were distributed among three locations (hereafter north, central, and south locations) spanning a latitudinal gradient in western Minnesota, USA (Figure 1). Soils of the south are glacial till, the north are laucustrine, and the central has regions containing both. Forty plots were located on conservation grasslands managed by the Minnesota Department of Natural Resources (DNR), eight plots managed by the US Fish and Wildlife Service, and 12 plots managed by private landowners as part of the CRP. Each plot was about 8 ha (20 acres; mean = 8.1 ha, SD = 0.5 ha) in size and contained a mixture of grasses and forbs. All plots were established more than five years prior to the project start date. Three of 12 CRP plots were planted with perennial introduced grasses and legumes (CP1) and the rest with perennial native grasses (CP2). The DNR plots were established with different species, but all were categorized as "restored/planted tall grass prairie". A list of the most frequently observed species is in Table S1. Plots were managed periodically for woody species with prescribed fire and/or mechanical harvest prior to the project start date. Fire was not implemented on our plots during the duration of the study. Occasional spot-spraying of herbicides was done in the south location to control invasive species.

Within each location, treatments were replicated in four blocks (Figure 1). Each block contained a control (no harvest) and three harvested plots. Since the control plots were not harvested, this analysis does not include data from those plots. Plots were randomly assigned a harvest treatment, and, for this analysis, were considered either a high- or low-intensity harvest. High-intensity treatments involved a complete harvest of the assigned plot while low-intensity treatments involved a partial harvest so that the plot contained a refuge of standing vegetation of 2 or 4 ha. The harvest treatments were designed to maintain other uses of the grassland, such as habitat for wildlife. In low-intensity harvest treatments, the refuge moved annually within the fixed plot area so that each year, a portion of the harvested area contained biomass that was not harvested the previous year. At all three locations, each block included one control plot, one high-intensity treatment, and two low-intensity treatments with refuges of 2 ha. A separate sub-study allowed the establishment of extra plots in the south location. Blocks in the south location included one extra high-intensity treatment plot and two extra low-intensity treatment plots (totaling seven plots per block). The extra low-intensity treatment plots had refuges of 4 ha. Twenty four plots were scheduled to be harvested in the south and twelve in each the central and north locations. Weather prevented the harvest of certain plots each year. No plots were harvested in the north in 2011 due to expiring land contracts.

5.2.2 FIELD AND LABORATORY METHODS

A single operator harvested the plots between late October and mid December in 2009, 2010, and 2011. No plots were harvested after the first significant snowfall. Vegetation was harvested to a target height of 15 cm with a self-propelled windrower with a mounted disc cutter. When conditions were deemed dry enough by the operator, the cut biomass was immediately baled using a large round baler. If the cut biomass required drying, it was raked into larger windrows and left to dry before being baled. Due to time constraints and landowner regulations, bales were removed from the plots as soon as possible, therefore individual bales were not weighed from each plot. Instead, bales were loaded onto semi trailers and weighed with a scale certified by the U.S. Department of Transportation on transport for storage. This weight was divided by the number of bales on the trailer to determine an average bale weight and variation (coefficient of variation = 9%; for further details, see Text S1). We divided the sum of all the trailer weights by the total number of bales to generate an overall average bale weight. The average bale weight was multiplied by the number of bales from each plot to estimate total harvested biomass. The perimeter of the cut area in each plot was measured using a hand-held global positioning system (GPS) (Garmin Ltd., Olathe, Kansas, USA) on an all-terrain vehicle. Biomass yield was determined for each plot as the amount of biomass harvested (Mg) divided by the area cut (ha).

While bales were still in the field, core samples were extracted from bales of harvested biomass for each plot with a hay probe (Forageurs Corp., Lakeville, MN, USA) attached to an electric drill. One biomass core was collected from every other bale as they were ejected from the baler; therefore the number of core samples was determined by the size of the harvested area within the plot and biomass productivity (mean number of cores in high-intensity plots = 22). Cores were aggregated by plot and weighed wet immediately after collection (mean sample weight = 156 g), dried at 45° C for four days, reweighed and used here to estimate bale yields on a dry matter basis.

Chemical constituents of the biomass were measured from the aggregated core samples for each plot. Biomass samples were dried at 45° C for four days, ground with a Wiley mill (Thomas-Wiley Mill Co., Philadelphia, PA, USA) to pass a 1 mm screen, and then reground with a cyclone mill. A subsample from each plot was analyzed for N by AgVise Laboratories using methods described on their website (Agvise Inc., Benson MN; http://www.agvise.com).

The concentration of cell wall carbohydrates was determined using near infrared spectroscopy (NIRS) with methods described by Schmer et al. [26]. NIRS estimates were from equations built with samples from previous collections, upon which wet chemistry methods were used to directly determine cell wall carbohydrate concentrations (Table S2). The values of xylose, arabinose, mannose, galactose, and glucose were calculated with methods established by the U.S. Department of Energy to predict theoretical ethanol conversion efficiency (Equation S1, http://www1.eere.energy.gov/biomass/ethanol_yield_calculator.html). Calculations used to estimate theoretical ethanol conversion efficiency assume 100% conversion efficiency because realized efficiency rates are not available for production-scale systems. In the summer of 2009, soil cores were collected to a depth of 20 cm at eight points adjacent to the randomly distributed vegetation quadrats. Soil cores were aggregated by plot and processed and analyzed by AgVise Laboratories for $N–NO_3$, pH, organic matter, and cation exchange capacity.

Plant community composition was visually assessed in 1.0×1.5 m quadrats at 12 random points within each plot in late July and/or early August of 2010 and 2011. A total of 24 quadrats were sampled in the high-intensity treatment plots in 2010 to assess sample power. In 2009, plant community data was collected from quadrats, each 0.75×5 m, in all plots. Quadrat locations were generated with ArcGIS 9.3 (ESRI, Redlands, CA, USA) and loaded to hand-held GPS units. Within each quadrat, surveyors identified all plant species and assigned each a score for relative abundance as a percentage of the canopy cover in the quadrat. Bare ground and litter were also assigned a percentage. Species were aggregated into functional groups for analysis. The average cover value for each functional group was calculated by plot.

Cooperative Farming Agreements, Special Use Permits, and a letter of approval were acquired from the Minnesota Department of Natural Resources, US Fish and Wildlife Service, and the US Department of Agriculture Farm Service Agency for permission to conduct research on state, federal and private land.

5.2.3 DATA ANALYSIS

Three response variables related to different components of bioenergy potential were measured in all plots and modeled in this study: biomass yield, theoretical ethanol conversion efficiency, and plant tissue N. Linear mixed effects models were used to test the main effect of location on the three response variables and to determine which covariates were significantly correlated with them. Total variation for each response variable was partitioned into four levels of a temporal/spatial hierarchy that was used as the random structure for the variance components analysis. The largest level of this hierarchy partitioned variance among years, with lower levels partitioning variance between locations, between blocks, and within plots; each level nested within the higher level. A model with only random effects was used to determine the variance at each level of the hierarchical random structure for all three response variables. Equation 1 was modified from West et al. [27] to derive variance estimates for each level of the random hierarchy, where ICCi represents the proportion of variation at level i compared with the total variation.

$$ICC_{Date} = \frac{\sigma^2_{Date}}{\sigma^2_{Date} + \sigma^2_{Location} + \sigma^2_{Block} + \sigma^2}$$

$$ICC_{Location} = \frac{\sigma^2_{Location}}{\sigma^2_{Date} + \sigma^2_{Location} + \sigma^2_{Block} + \sigma^2}$$

$$ICC_{Block} = \frac{\sigma^2_{Block}}{\sigma^2_{Date} + \sigma^2_{Location} + \sigma^2_{Block} + \sigma^2}$$

To quantify the differences in biomass yield, ethanol conversion efficiency, and plant N between locations, a dummy variable was assigned to the south, central, and north locations and was modeled as a categorical main fixed effect. Using location as a fixed effect, various random structures composed of the nested spatial/temporal variables were fit to models and compared using maximum likelihood ratio tests.

Land ethanol yield (1 ha⁻¹) was calculated by multiplying ethanol conversion efficiency (1 Mg⁻¹) by biomass yield (Mg ha⁻¹) for each plot. A linear regression model was used to estimate the fraction of variation in land ethanol yield due to variation in biomass yield.

For each response variable, we selected a group of candidate covariates a priori from a list of measured variables (Table 1). A global model for each re-

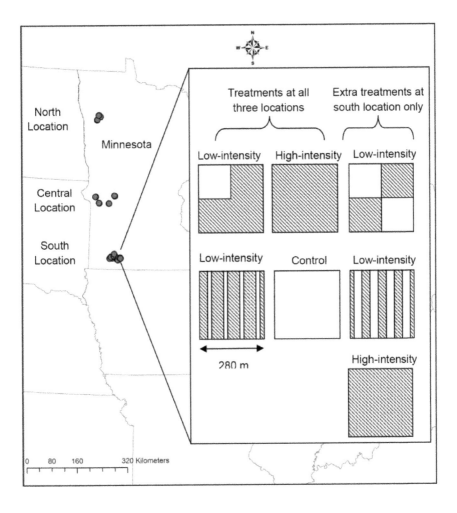

FIGURE 1: Study areas in Minnesota, located in the Upper Midwest, USA. Research blocks are indicated by circles within the outline of Minnesota in north, central, and south locations. Inset outlines treatments within blocks.

sponse variable included all covariates related to plant community structure and an interaction between each community covariate and the main effect of location. No three-way interactions were tested. Each global model included a best fitting random structure and a first order autocorrelation structure. The global model was reduced by removing the least significant fixed effect determined by t-statistic at P<0.05 [28]. This iterative process continued until all fixed effects were removed. The resulting models were compared using Akaike's information criteria adjusted for small sample sizes (AICc) [29]. The best fitting model was refit using restricted maximum likelihood to generate unbiased parameter estimates. For models without interactions, Tukey's post hoc means separation test was used to determine differences between levels of significant main effects.

TABLE 1: List and description of all covariates available for analysis.

Effect	Variable	Description
Random	DATE, LOC, BLOCK, PLOT	Nested temporal and spatial variables. Plot nested in block nested in location.
Main	Location	Categorical main effects of location.
Plant Community plot.	C4, C3, Legume, Forb	Continuous measure of mean percent cover of each plant functional group by
Soil Fertility	NO₃, OM, pH, CEC	Mean values of N-N03 (N03), organic matter (OM), pH, and cation exchange capacity (CEC) by plot.
Plant Composition	PlantN	The concentration of N in harvested biomass tissue.
Precipitation	April, May, June, July, August, September	Total monthly precipitation measured for each year by block.
Interactions	C4 x Location, C3 x Location, Legume x Location, Forb x Location, Harvest x Location	Interaction between main effects, and between the main effect of location and all plant community covariates

A mixed effect model was used to test the effect of harvest intensity on the change in biomass yield over time. The difference in biomass yield from the first harvest (2009) to the last (2011) was calculated for plots in the south and central locations to test the hypothesis that trends in biomass yields through time would be the same for plots where all the biomass is removed as plots that include regions of previously unharvested biomass. The change in yield was compared between low- and high-intensity harvest treatments. The model included an interaction between harvest intensity and location while accounting for variation in each plot as a random variable. All statistical analyses were conducted with program R [30].

5.3 RESULTS

We analyzed and modeled biomass yield from 109 observations and theoretical ethanol conversion efficiency and plant tissue N from 112 observations from conservation grasslands harvested in autumn of 2009, 2010, and 2011. Weather obstructed biomass harvest at certain plots each year which resulted in an unbalanced data set. No plots were harvested in the north location in 2011 due to expiring land contracts.

The south location received more precipitation during the growing season compared with the north and central locations during all years of the study. Precipitation was lowest in 2009 at the south and central locations, and lowest in 2011 at the north. Over the course of the project, precipitation was the greatest in 2010 and well exceeded the 30-year mean at all locations. In 2011, the north and central locations were below the 30-year mean while precipitation at the central location was higher (Table 2).

5.3.1 BIOMASS YIELD

Without accounting for covariates, mean biomass yield in the south was 55%, 69%, and 55% greater than other locations in 2009, 2010, and 2011 respectively (Figure 2A). Annual plot biomass yield ranged from 0.5 Mg ha^{-1} to 5.7 Mg ha^{-1} and had an overall mean of 2.5 Mg ha^{-1} across all locations and years. Biomass yield increased from 2009 to 2011 in both the south and central locations and in both harvest intensities (Figure 3). The increase in biomass yield through time was the same between harvest intensities (F = 0.48, df = 27, P = 0.49).

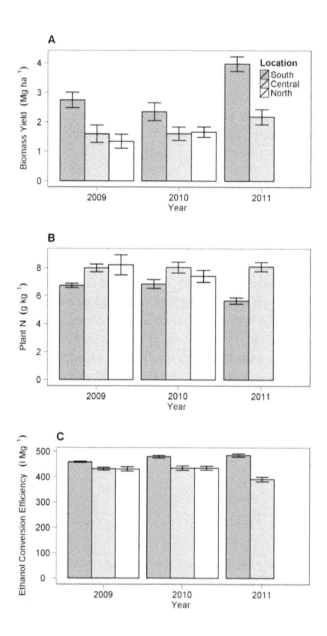

FIGURE 2: Average values (SE) of response variables by location and year. Mean values of biomass yield (A), plant tissue N (B), and ethanol conversion efficiency (C). Black, gray and white bars are mean values from plots harvested in south, central and north locations respectively.

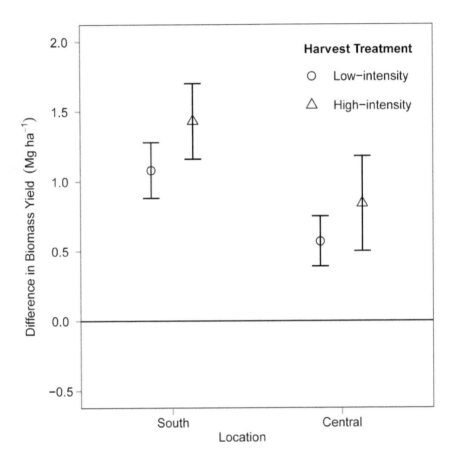

FIGURE 3: Change in biomass yield from 2009 to 2011 in low- and high-intensity harvest treatments by location. Average change in biomass yield(±90% CI). In low-intensity plots, one third to one half of the annually harvested biomass was from an area not previously harvested. High-intensity harvest plots included biomass from the same area harvested annually.

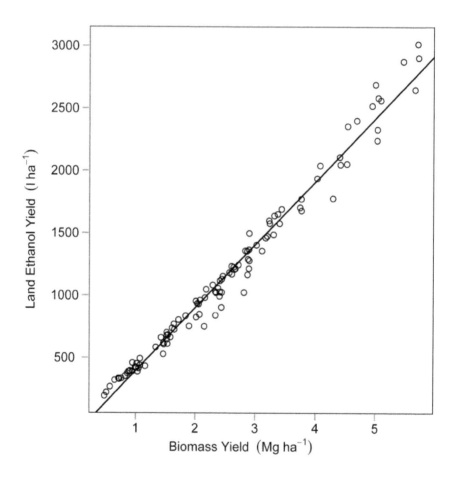

FIGURE 4: Correlation between land ethanol yield (l ha⁻¹) and biomass yield (Mg ha⁻¹). Points represent values from conservation grasslands harvested in the autumn of 2009, 2010, and 2011. Regression line from linear model with R-squared value = 0.98.

TABLE 2: Cumulative precipitation from April through October by location and year, for comparison with other regions.

	2009	2010	2011	30 yr. mean
	(mm)			
North	435	663.46	391.51	442.21
Central	452.64	663.22	538.59	518.92
South	559.09	864.36	577.13	582.93

30 yr mean: http://hurricane.ncdc.noaa.gov/climatenormals/clim81/MNnorm.pdf. Minnesota Climatology Working Group: http://climate.umn.edu/ hidradius/HIDENbrowse_ PHP.asp

5.3.2 BIOMASS QUALITY

Biomass yield was a significant predictor of the variation in land ethanol yield (F = 5558, df = 1 and 108, P<0.001). The adjusted R-squared was 0.98 for the relationship between biomass yield and land ethanol yield (Figure 4). Mean ethanol conversion efficiency was 450 l Mg^{-1} with a standard deviation of 38 across all locations and years. Mean plant N concentration was 7.1 g kg^{-1} with a standard deviation of 1.5 and was not consistently different among locations and years. Mean plant N was lower and mean ethanol conversion efficiency was greater in the south than the other locations in all three years (Figure 2B and 2C).

5.3.3 VARIANCE COMPONENTS ANALYSIS

Results from the intercept-only random effects models suggest that of the total variation in biomass yield, ethanol conversion efficiency, and plant N, the variance between years explained the smallest fraction (Table 3). The largest fraction of the variance in biomass yield and plant N was partitioned into within-plot variance, while the variation between locations accounted for about one-third for both responses. More than a majority of variation in ethanol conversion efficiency was observed between locations (Table 3).

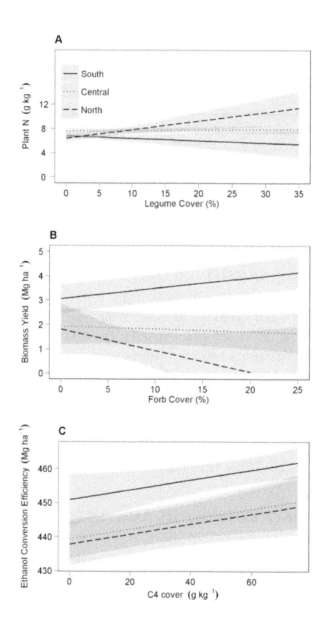

FIGURE 5: Estimated effect of plant functional group composition on bioenergy potential. Regression line estimates(±90% CI) of the effect of legume cover on the concentration of N in biomass after harvest (A), the effect of forb cover on biomass yield (B), and the effect of C4 cover on ethanol conversion efficiency (C). Estimates are from the best fitting models with all other covariates held constant at their average values.

TABLE 3: The contribution of variation from nested random effects for measures of bioenergy quantity and quality.

Nested Sources of Variation	Biomass Yield	Ethanol Conversion Efficiency	Plant N
Between years	0.33 (6%)	$4.6*10^{-3}$ (0%)	$1.0*10^{-4}$ (0%)
Between locations	0.74 (31%)	28.78 (57%)	0.86 (34%)
Between blocks	0.65 (24%)	17.45 (21%)	0.15 (1%)
Within plot (residual)	0.82 (39%)	17.85 (22%)	1.18 (65%)

Variation reported as standard deviation and percent of total variation.

5.3.4 BIOENERGY POTENTIAL MODELS

5.3.4.1 BIOMASS YIELD.

Measured soil fertility variables did not contribute to explained variation in biomass yield. The effect of forb cover was significant in the best fitting model (Table 4) and influenced biomass yield uniquely in the south compared with the other locations (Table 5, Figure 5B). Specifically, forb cover was negatively correlated with biomass yield in the central and north locations, but positively correlated with biomass yield in the south location. Covariates for May precipitation and legume cover were positively correlated with biomass yield in the best fitting model (Table 5). A model with the random variables plot (identified below as PLOT; see Table 1) nested within block (identified as BLOCK) was superior to a model without random effects (L = 40.77, df = 1, P<0.001). The three best fitting models were similar in their explanatory power determined by AICc (Table 4).

5.3.4.2 ETHANOL CONVERSION EFFICIENCY.

The two best fitting models included the effect of location, the cover of C4 grass, and the nitrogen content of harvested biomass as predictors of variation in ethanol conversion efficiency. The best fitting model included the cover of forbs and omitted all interactions between main effect and covariates (Table 4). The cover of C4 grass was positively correlated with ethanol conversion efficiency (Figure 5C), while plant N and forb cover showed negative relationships with ethanol conversion efficiency (Table 5). Ethanol conversion efficiency was significantly greater in the south than the central (P = 0.034) and north (P = 0.020) locations, with a metric ton of biomass producing 12% more ethanol in the south than the average of the central and north locations. There was no significant difference between the central and north (P = 0.947) locations. A model with the random variables BLOCK and DATE was best supported for explaining variation in ethanol conversion efficiency. The random structure was fit to allow unique BLOCK variation around the intercept by DATE. This structure was better supported than the fully nested random structure (L = 13.5, df = 1, P = 0.004) and a model without a random structure (L = 64.7, df = 1, P<0.001). The two best fitting models differed by 0.69 AICc points and one parameter (Table 4).

TABLE 4: Top three best-supported models of bioenergy potential measured from conservation grasslands in Minnesota, USA.

Response	Model	Parameters (K)	ΔAICc
Biomass Yield	Intercept+Location x Forb+May+Legume	12	0.00
	Intercept+Location x Forb+Legurne+May+June	13	1.56
	Intercept+Location x Forb+Forb+May	10	2.06
Ethanol conversion efficiency	Intercept+Location+C4+PlantN+Forb	14	0.00
	Intercept+Location+C4+PlantN	13	0.69
	Intercept+Location+C4+Forb+N03+PlantN	15	1.86
Plant N	Intercept+Location x Legurne+C4+N03	12	0.00
	Intercept+Location x Legurne+C4+N03 +pH	13	0.28
	Intercept+Location+C4+N03	9	0.42

5.3.4.3 PLANT N.

The three best fitting models included the main effect of location, C_4 cover, and soil $N–NO_3$ concentration (Table 4). The best supported model included an interaction term between location and legume cover (Table 5). In the south, legume cover was negatively correlated with plant N as opposed to the positive correlation observed in the central and north locations (Figure 5A). Soil $N-NO_3$ and C_4 cover were positively and negatively correlated with plant N respectively (Table 5). The best fitting random structure for modeling the concentration of N in biomass included PLOT nested within BLOCK. This structure was superior to a model without a random component (L = 14.9, df = 1, P<0.001) and to a model with a fully nested hierarchy of random variables (L = 9.2, df = 1, P = 0.003).

TABLE 5: Parameter estimates from best-fitted mixed effects models with biomass yield, ethanol conversion efficiency, and plant N as response variables.

Response	Variable	β	SE (β)	DF	t-value	p-value
Biomass Yield	Intercept	2.069	0.381	56	5.432	< 0.001
	Location 2	-1.126	0.583	9	-1.932	0.085
	Location 3	-1.243	0.738	9	-1.684	0.126
	May	0.011	0.001	56	9.893	< 0.001
	Legume	0.017	0.007	56	2.428	0.018
	Forb	0.044	0.013	56	3.284	0.002
	Location 2 x Forb	-0.055	0.026	56	-2.073	0.043
	Location 3 x Forb	-0.132	0.076	56	-1.750	0.086
Ethanol Conversion	Efficiency Intercept	529.905	9.680	96	54.743	< 0.001
	Location 2	-11.550	4.623	9	-2.498	0.034
	Location 3	-13.005	4.840	9	-2.687	0.025
	C4	0.147	0.070	96	2.081	0.040
	Plant N	-10.812	1.088	96	-9.941	< 0.001
	Forb	-0.357	0.203	96	-1.760	0.082
Plant N	Intercept	6.786	0.458	59	14.827	< 0.001
	Location 2	0.746	0.400	9	1.862	0.096
	Location 3	-0.384	0.531	9	-0.724	0.488
	C4	-0.017	0.006	59	-2.975	0.004
	Legume	-0.040	0.043	59	-0.925	0.359
	NO_3	0.077	0.016	59	4.748	< 0.001
	Location2 x Legume	0.050	0.044	59	1.137	0.260
	Location3 x Legume	0.182	0.071	59	2.579	0.012

5.4 DISCUSSION

Harvested biomass yields from low-input grasslands managed for conservation was 2.5 Mg ha^{-1} and on average, fluctuated 23% around this mean across the three year study period. Assuming this yield can be achieved from all the conservation grasslands within an 80 km radius of a biorefinery located in the southwest portion of Minnesota (a total of 107,571 ha of conservation grassland or 5.4% of the total area), and that only 75% of the conservation grasslands are harvestable within that area, approximately 1000 Gw*hours of energy is available (Text S2). If divided across the year, this is equivalent to 114 MW of continuous energy from conservation grasslands alone.

Yields were highest in the south location in all years of this experiment, but were 49% lower than first-year hand-cut yield estimates from newly established high diversity mixtures grown in similar regions [31]. Despite similar growing conditions, the high diversity mixtures were grown on fine loam soil with N, P, and K concentrations more than two times higher than concentrations found in our soils. From our southern plots, biomass yield estimates from hand-cut samples collected in late July were 91% and 54% greater than yield values from commercial-scale harvest in 2010 and 2011 respectively (unpublished data), both of which are similar to the harvest efficiency of managed switchgrass plots in Italy [32]. Although leaf loss and reallocation of C to belowground structures can account for 12% to 19% of decreased biomass yields from September to November [33], there is evidence that commercial-scale harvesting techniques can be made more efficient at both cutting more of the material to a desired height and picking up more of the material with a baler to improve yields [32]. It should be noted that stubble and residual litter provides environmental benefits by reducing erosion and providing cover for ground nesting birds, therefore 100% harvest efficiency may not be a desired objective. Observed variation in litter quantities across studies suggests that caution be taken when comparing aboveground productivity estimates and biomass yields between small-scale and large-scale studies that do not use similar cutting and biomass collection methods.

Generally, the concentration of N in herbaceous biomass results in greater NO_x emissions during thermochemical conversion to energy compared with light fuel oil and natural gas [34]. It has been recommended to delay harvesting until after senescence to allow perennial plants to translocate N to belowground tissues for both switchgrass [35] and conservation grassland biomass [16]. Nitrogen content in harvested biomass from this project was similar to conservation grasslands harvested after a killing frost in South Dakota [36]. There is concern that low-input grasslands might not be a long-term viable source of biomass because of N depletion during harvest [37], but those concerns have not yet been tested. There is evidence that long-term annual biomass harvest from low-input grasslands does not decrease yields [38]. Mixed-species grasslands like those used in this project contain legumes that add N annually. N inputs via legumes ranged from 28 to 187 kg ha^{-1} in mowed grass/legume pastures that contained white clover [39], yet studies are needed to determine the net N flux in harvested grassland systems across a range of locations.

Variation in biomass yield, ethanol conversion efficiency, and concentration of N in plant tissue was relatively small between years, deviating from each location's average by no more than +/− 27%, 11%, and 7% respectively. This is in contrast to other studies with less mature perennial grasslands (our study sites were all >5years old), where issues with establishment contributed to larger (up to 69%) year-to-year variation in biomass yield [21]. Across the total study area, between-year variability in biomass yield was small despite differences in precipitation. Our results show that precipitation during the month of May measured at the block level is important in determining biomass yield (Figure 6). Total precipitation may not be a good indicator for predicting biomass yields because high amounts of precipitation during harvesting months may result in lower yields due to leaf losses and other inefficiencies in biomass collection, especially when harvesting with production-scale equipment [32]. Excessive precipitation during autumn months inundated some parts of this experiment and prevented the harvest of certain plots each year. Averaged across all years, 83%, 78%, and 74% of the planned harvested areas were harvested in the south, central and north locations respectively. This percentage increased annually in the south and central locations.

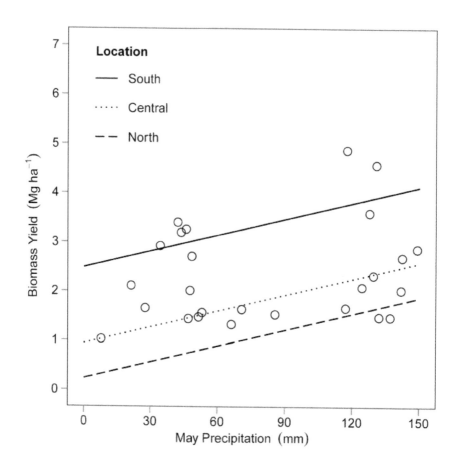

FIGURE 6: Estimated effect of May precipitation on biomass yield. Dots represent average measured biomass yield and May precipitation values by block. Regression lines are model estimates for bioenergy yield across the precipitation gradient for each location, with all other covariates held constant at their average values.

Consistent values for biomass quality metrics are important for viable biorefinery production. A substantial fraction of the total variation in biomass yield was observed between locations, which is in accordance with studies on the variation of switchgrass yield [21]. About one-quarter of the total variation in biomass yield was measured between blocks, which was similar to the results of yield variation in C3-dominated grasslands analyzed for bioenergy [20]. Florine et al. reported smaller total variation in plant N (SD = 0.4 g kg^{-1}) than our results (SD = 1.5 g kg^{-1}) [20]. Total variation in ethanol conversion efficiency was relatively small but greater than reports from switchgrass, yet similar in terms of partitioning between spatial and temporal scales [26].

The variation in land ethanol yield was almost exclusively due to variation in biomass yield (Figure 4). Land managers looking to harvest biomass from conservation grassland for ethanol production would maximize revenues by identify high biomass yielding plots as opposed to harvesting plots based on the theoretical ethanol potential of the plants.

We hypothesized that covariates would explain variation among locations (Table 6). However, for all response variables, location remained a significant variable in the best fitting models (Table 5). Best fitting models for biomass yield and plant N included interactions between location and plant community covariates, which provide limited information to draw conclusions as to why differences in these response variables exist across locations. In terms of ethanol conversion efficiency, location was identified as a main source of variation, therefore suggesting that other factors related to space–factors that were not measured in this study–influenced the response.

Other reports have suggested that plant community characteristics such as C4 grass cover [14] and planted species richness [2] improve biomass yields. In this study, it was the cover of non-legume forbs that explained variation in biomass yield (Table 4 and 5). In the south location, plots with greater average forb cover had higher biomass yields, while in the central and north locations, increasing forb cover was associated with lower yields. We expected, as Adler et al. documented, that the cover of C4 grass would be positively correlated with biomass yield, and our competitive models include that variable (Table 4). It is possible that an increase in forb cover displaces C4 grasses, which would explain the negative correlation between forb cover and biomass yield in the central and north locations. The inverse relationship between forb cover and biomass yield in the south could be driven by a high-yielding forb species that is present or abundant in the south but not in the other locations. We explored this possibility and found that common milkweed (*Asclepias syriaca*) was present in 300 sample points in the south and only 50 and 5 sample points in the central and north locations. Using data from all sample points, a Pearson's correlation test showed that the cover of common milkweed was not correlated to the cover of C4 grass ($P = 0.303$) but was correlated to biomass yield ($P = 0.016$). This suggests that common milkweed could increase biomass yield without displacing C4 grass cover (Table 6). Other studies have observed increases in forb abundance without associated decreases in biomass production [40].

TABLE 6: Mean values (SD) of covariates by location across all years from conservation grasslands in Minnesota.

Covariate	South	Central	North
		% cover	
C4	56.86 (18.78)	24.94 (18.37)	20.12 (18.71)
C3	18.15 (16.30)	37.77 (19.58)	45.64 (23.15)
Legume	2.80 (3.22)	8.51 (14.57)	4.81 (5.07)
Forb	6.54 (6.57)	10.35 (5.94)	6.26 (3.22)
NO$_3$	7.84 (3.94)	11.04 (8.35)	13.76 (12.22)
OM	5.27 (1.33)	6.52 (3.04)	5.38 (1.65)
pH	6.67 (0.49)	7.52 (0.37)	7.68 (0.65)
CEC	22.17 (7.55)	25.66 (7.44)	26.19 (8.08)

Harvested areas in the low-intensity harvest treatments included a fraction of the plot where vegetation was left standing the year before. This did not affect biomass yields compared with completely harvested plots. European mixed-species hay yields did not decrease after decades of annual harvest without nutrient inputs [38], though long term studies are needed to verify if similar patterns exist in North American grasslands. The positive correlation of May precipitation with yield could be because it supplies resources before the peak productivity time of C4 grasses, which contribute to biomass yield when harvested in autumn [36]. Other studies have shown that the variation in June soil moisture was positively correlated with C4 grass productivity [41], but soil moisture measurements were not made in our study.

Maximum theoretical ethanol conversion efficiency values were slightly higher than those reported in switchgrass [26] and similar to mixed prairies [42], and were greater in biomass harvested from the south compared with biomass from the central and north locations (Figure 2C). Studies of switchgrass show that harvesting later after plant senescence results in higher potential ethanol conversion efficiency [43], thus a similar pattern could exist in polyculture grasslands. We harvested plots in sequence from the north to the south so that the plants would be at a similar phenological stage at the time of cutting. A negative correlation between plant tissue N and ethanol conversion efficiency was apparent in this study (Table 5), and since plant N decreases with senescence, the later harvest date in the south location may have contributed to higher ethanol conversion efficiency found here. Also, our results confirm previous reports of correlations between C4 grass cover and ethanol conversion efficiency [14] (Figure 5C). In general, C4 grasses have higher levels of fermentable sugars than forbs [44]; therefore ethanol conversion efficiency is expected to decrease with increased forb cover relative to C4 dominated stands. As highlighted in this study, Gillitzer et al. showed that the relationship between species composition and biomass yield, rather than species composition and ethanol conversion efficiency, is the more dominant driver of land ethanol yield [42], [45].

Legumes in mixed-species grasslands fix atmospheric nitrogen, which has several consequences for ecosystem functioning including increased productivity [46]. However, in the case of combustion bioenergy, undesirable consequences of legume biomass come in the form of pollution. Legume biomass has relatively higher levels of tissue N than forbs and grasses, which can lead to greater NO_x emissions during thermochemical energy conversion [34]. The best fitting model identified a relatively strong trend in legume cover and plant N in the north location (t = 2.579, P = 0.012). Weaker evidence of a relationship was observed in the central (t = 1.137, P = 0.260) and the south locations (t = −0.925, P = 0.359), which could be related to the absence or presence of a specific legume species, as observed in other studies [47]. The estimates from this model predict that a four-fold increase in legume cover (from the observed average of 4.8% to 19.2%) in the north location would increase biomass N concentrations approximately 23%, or to a value of 10.2 g N kg^{-1}. Promoting legumes increases functional group diversity, which leads to other ecological benefits including increased soil carbon storage [48]. Also, complementarity among C4 grasses and legumes increases biomass yields [48]. Therefore, we believe that the model-estimated environmental cost of legume abundance in bioenergy grasslands is far outweighed by the ecological and yield benefits they provide.

The three best supported models all suggest that unfertilized soils with naturally higher levels of N–NO$_3$ will produce biomass with greater concentrations of tissue N (Table 4). Elevated levels of soil N–NO$_3$ could come as a result of N fertilizer, which has been considered as a management tool to increase biomass yields in conservation grasslands [8], [23]. Fertilization experiments show that higher N fertilizer rates lead to higher concentrations of N in biomass tissue for C3-dominated mixed grasslands [49], for switchgrass [50], and other C4 grasses [51]. Nitrogen fertilization can lead to a loss of species and functional group turnover [52], but when fertilized grasslands are harvested, species diversity has been shown to be maintained [53] or increase [40]. When considering N fertilizers, land managers must weigh the potential benefits for biomass yields against potential detrimental effects including undesirable shifts in species composition and decreased biomass quality.

5.5 CONCLUSIONS

Biomass quality from mixed-species grasslands not managed for bioenergy is similar to dedicated energy feedstocks, in terms of theoretical ethanol conversion efficiency and biomass N. Almost all of the variation in land ethanol yield is based on biomass yield, therefore efforts should be focused on maximizing biomass yield rather than biomass quality when managing grasslands for land ethanol yield. A combination of climate, soil fertility, and plant community factors influence overall bioenergy potential. The effect of forbs and legumes on biomass yield and tissue N, respectively, were different in the south compared with the central and north locations. The covariates we measured did not explain why theoretical ethanol conversion efficiency was greater in the south compared with the other locations, but the cover of C4 grass was positively correlated with ethanol conversion efficiency. After three continuous years of harvest, leaving a portion of standing biomass within the harvested area does not influence biomass yield of future harvests. Simply focusing on plant community variables to predict bioenergy potential of conservation grasslands across various locations at the scale we studied will not provide accurate estimates; instead attention should be drawn to local variation in soil fertility, climate, and possibly plant species and interactions between these variables.

REFERENCES

1. Fargione J, Hill J, Tilman D, Polasky S, Hawthorne P (2008) Land clearing and the biofuel carbon debt. Science 319: 1235–1238. doi: 10.1126/science.1152747
2. Tilman D, Hill J, Lehman C (2006) Carbon-negative biofuels from low-input high-diversity grassland biomass. Science 314: 1598–1600. doi: 10.1126/science.1133306
3. Robertson BA, Doran PJ, Loomis LR, Robertson JR, Schemske DW (2010) Perennial biomass feedstocks enhance avian diversity. GCB Bioenergy 3: 235–246. doi: 10.1111/j.1757-1707.2010.01080.x
4. Meehan TD, Hurlbert AH, Gratton C (2010) Bird communities in future bioenergy landscapes of the Upper Midwest. Proc Natl Acad Sci USA 107: 18533–18538. doi: 10.1073/pnas.1008475107
5. Gardiner MA, Tuell JK, Isaacs R, Gibbs J, Ascher JS, et al. (2010) Implications of three biofuel crops for beneficial arthropods in agricultural landscapes. Bioenergy Res 3: 6–19. doi: 10.1007/s12155-009-9065-7
6. Glover JD, Culman SW, DuPont ST, Broussard W, Young L, et al. (2010) Harvested perennial grasslands provide ecological benchmarks for agricultural sustainability. Agric Ecosyst Environ 137: 3–12 doi:10.1016/j.agee.2009.11.001.
7. Johnson MV, Kiniry JR, Sanchez H, Polley HW, Fay PA (2010) Comparing Biomass yields of low-input high-diversity communities with managed monocultures across the central United States. Bioenergy Res 3: 353–361 doi:10.1007/s12155-010-9094-2.
8. Lee D, Aberle E, Chen C, Egenolf J, Harmoney K, et al. (2013) Nitrogen and harvest management of Conservation Reserve Program (CRP) grassland for sustainable biomass feedstock production. GCB Bioenergy 5: 6–15 doi:10.1111/j.1757-1707.2012.01177.x.

9. Gelfand I, Zenone T, Jasrotia P, Chen J, Hamilton SK (2011) Carbon debt of Conservation Reserve Program (CRP) grasslands converted to bioenergy production. Proc Natl Acad Sci USA 108: 13864–13869 doi:10.1073/pnas.1017277108/-/DCSupplemental www.pnas.org/cgi/doi/10.1073/pnas.1017277108.

10. Hill J, Nelson E, Tilman D, Polasky S, Tiffany D (2006) Environmental, economic, and energetic costs and benefits of biodiesel and ethanol biofuels. Proc Natl Acad Sci USA 103: 11206–11210 doi:10.1073/pnas.0604600103.

11. Olson D (2007) Sustainable biomass land reserves for a sustainable future. Minneapolis: Institute for Agriculture and Trade Policy..4 p

12. Wiens J, Fargione J, Hill J (2011) Biofuels and biodiversity. Ecol Appl 21: 1085–1095. doi: 10.1890/09-0673.1

13. Jungers JM, Lehman CL, Sheaffer CC, Wyse DL (2011) Characterizing grassland biomass for energy production and habitat in Minnesota. Proceedings of the 22nd North American Prairie Conference: 168–171.

14. Cai X, Zhang X, Wang D (2011) Land availability for biofuel production. Environ Sci Technol 45: 334–339 doi:10.1021/es103338e.

15. Adler PR, Sanderson MA, Weimer PJ, Vogel KP (2009) Plant species composition and biofuel yields of conservation grasslands. Ecol Appl 19: 2202–2209. doi: 10.1890/07-2094.1

16. Venuto BC, Daniel JA (2010) Biomass feedstock harvest from Conservation Reserve Program land in northwestern Oklahoma. Crop Sci 50: 737–743 doi:10.2135/cropsci2008.11.0641.

17. Baxter LL, Miles TR, Jenkins BM, Milne T, Dayton D, et al. (1998) The behavior of inorganic material in biomass-fired power boilers: field and laboratory experiences. Fuel Processing Technology 54: 47–78 doi:10.1016/S0378-3820(97)00060-X.

18. Robertson GP, Hamilton SK, Del Grosso SJ, Parton WJ (2011) The biogeochemistry of bioenergy landscapes: carbon, nitrogen, and water considerations. Ecol Appl 21: 1055–1067. doi: 10.1890/09-0456.1

19. David K, Ragauskas AJ (2010) Switchgrass as an energy crop for biofuel production: A review of its ligno-cellulosic chemical properties. Energy Environ Science 3: 1182 doi:10.1039/b926617h.

20. Florine S, Moore K, Fales S, White T, Leeburras C (2006) Yield and composition of herbaceous biomass harvested from naturalized grassland in southern Iowa. Biomass Bioenergy 30: 522–528 doi:10.1016/j.biombioe.2005.12.007.

21. Schmer MR, Mitchell RB, Vogel KP, Schacht WH, Marx DB (2009) Spatial and temporal effects on wwitchgrass wtands and yield in the Great Plains. Bioenergy Res 3: 159–171 doi:10.1007/s12155-009-9045-y.

22. Divirgilio N, Monti A, Venturi G (2007) Spatial variability of switchgrass (Panicum virgatum L.) yield as related to soil parameters in a small field. Field Crops Res 101: 232–239 doi:10.1016/j.fcr.2006.11.009.

23. Mulkey VR, Owens VN, Lee DK (2006) Management of switchgrass-dominated Conservation Reserve Program lands for biomass production in South Dakota. Crop Sci 46: 712–720 doi:10.2135/cropsci2005.04-0007.

24. Lee DK, Owens VN, Doolittle JJ (2007) Switchgrass and soil carbon sequestration response to ammonium nitrate, manure, and harvest frequency on Conservation Reserve Program land. Agron J 99: 462–468 doi:10.2134/agronj2006.0152.

25. Jefferson PG, Wetter L, Wark B (1999) Quality of deferred forage from waterfowl nesting sites on the Canadian prairies. Can J Anim Sci 79: 485–490. doi: 10.4141/a98-123

26. Schmer MR, Vogel KP, Mitchell RB, Dien BS, Jung HG, et al. (2012) Temporal and Spatial Variation in Switchgrass Biomass Composition and Theoretical Ethanol Yield. Agron J 104: 54 doi:10.2134/agronj2011.0195.

27. West BT, Welch KB, Galecki AT (2007) Linear mixed models: A practical guide using statistical software. Boca Raton, FL: Taylor and Grancis Group. .359 p

28. Zuur AF, Ieno EN, Walker NJ, Saveliev AA, Smith GM (2010) Mixed effects models and extensions in ecology with R. New York: Springer. 574 pdoi:10.1007/978-0-387-87458-6.

29. Burnham KP, Anderson DR (2002) Model selection and multi-model inference: a practical information-theoretic approach. Second Edition. New York: Springer. 353 p.

30. R Development Core Team (2010) R: A language and evironment for statistical computing. Available: http://www.r-project.org/. Accessed 2013 Mar 14.

31. Mangan ME, Sheaffer C, Wyse DL, Ehlke NJ, Reich PB (2011) Native perennial grassland species for bioenergy: Establishment and biomass productivity. Agron J 103: 509–519 doi:10.2134/agronj2010.0360.

32. Monti A, Fazio S, Venturi G (2009) The discrepancy between plot and field yields: Harvest and storage losses of switchgrass. Biomass Bioenergy 33: 841–847. doi: 10.1016/j.biombioe.2009.01.006

33. Sanderson M, Read J, Reed R (1999) Harvest management of switchgrass for biomass feedstock and forage production. Agron J 91: 5–10. doi: 10.2134/agronj1999.00021962009100010002x

34. Nussbaumer T (2003) Combustion and co-combustion of biomass: Fundamentals, technologies, and primary measures for emission reduction. Energy Fuels17: 1510–1521. doi: 10.1021/ef030031q

35. Ogden CA, Ileleji KE, Johnson KD, Wang Q (2010) In-field direct combustion fuel property changes of switchgrass harvested from summer to fall. Fuel Processing Technology 91: 266–271 doi:10.1016/j.fuproc.2009.10.007.

36. Mulkey VR, Owens VN, Lee DK (2008) Management of warm-season grass mixtures for biomass production in South Dakota USA. Bioresour Technol 99: 609–617 doi:10.1016/j.biortech.2006.12.035.

37. Russelle MP, Morey RV, Baker JM, Porter PM, Jung HG (2007) Comment on "Carbon-negative biofuels from low-input hight-diversity grassland biomass.". Science 316: 1567. doi: 10.1126/science.1139388

38. Jenkinson DS, Potts JM, Perry JN, Barnett V, Coleman K, et al. (1994) Trends in herbage yields over the last century on the Rothamsted Long-term Continuous Hay Experiment. J Agric Sci 122: 365–374. doi: 10.1017/s0021859600067290

39. Jarchow ME, Liebman M (2012) Nitrogen fertilization increases diversity and productivity of prairie communities used for bioenergy. GCB Bioenergy. doi: 10.1111/j.1757-1707.2012.01186.x.

40. Ledgard S (2001) Nitrogen cycling in low input legume-based agriculture, with emphasis on legume/grass pastures. Plant Soil 228: 43–59.

41. Nippert JB, Knapp AK, Briggs JM (2006) Intra-annual rainfall variability and grassland productivity: Can the past predict the future? Plant Ecol 184: 65–74 doi:10.1007/s11258-005-9052-9.

42. Jarchow ME, Liebman M, Rawat V, Anex RP (2012) Functional group and fertilization affect the composition and bioenergy yields of prairie plants. GCB Bioenergy 4: 671–679 doi:10.1111/j.1757-1707.2012.01184.x.

43. Adler PR, Sanderson MA, Boateng AA, Weimer PJ, Jung HG (2006) Biomass yield and biofuel quality of switchgrass harvested in fall or spring. Agron J 98: 1518–1525 doi:10.2134/agronj2005.0351.

44. Lee D, Owens VN, Boe A, Jeranyama P (2007) Composition of herbaceous biomass feedstocks. Brookings, SD: North Central Sun Grant Center. 16 p.

45. Gillitzer PA, Wyse DL, Sheaffer DD, Lehman CL (2012) Biomass production potential of grasslands in the oak savanna region of Minnesota, USA. BioEnergy Res. doi: 10.1007/s12155-012-9233-z.

46. Tilman D, Knops JMH, Wedin D, Reich P, Ritchie M, et al. (1997) The influence of functional diversity and composition on ecosystem processes. Science 277: 1300–1302 doi:10.1126/science.277.5330.1300.

47. Spehn EM, Schmid B, Hector A, Caldeira MC, Dimitrakopoulos PG, et al. (2002) The role of legumes as a component of biodiversity in a cross-European study of grassland biomass nitrogen. Oikos 2: 205–218. doi: 10.1034/j.1600-0706.2002.980203.x

48. Fornara DA, Tilman D (2008) Plant functional composition influences rates of soil carbon and nitrogen accumulation. J Ecol 96: 314–322 doi:10.1111/j.1365-2745.2007.01345.x.

49. Malhi SS, Nyborg M, Soon YK (2010) Long-term effects of balanced fertilization on grass forage yield, quality and nutrient uptake, soil organic C and N, and some soil quality characteristics. Nutr Cycl Agroecosyst 86: 425–438 doi:10.1007/s10705-009-9306-3.

50. Guretzky JA, Biermacher JT, Cook BJ, Kering MK, Mosali J (2010) Switchgrass for forage and bioenergy: harvest and nitrogen rate effects on biomass yields and nutrient composition. Plant Soil 339: 69–81 doi:10.1007/s11104-010-0376-4.

51. Waramit N, Moore KJ, Heggenstaller AH (2011) Composition of native warm-season grasses for bioenergy production in response to nitrogen fertilization rate and harvest date. Agron J 103: 655 doi:10.2134/agronj2010.0374.

52. Suding KN, Collins SL, Gough L, Clark C, Cleland EE, et al. (2005) Functional- and abundance-based mechanisms explain diversity loss due to N fertilization. Proc Natl Acad Sci USA 102: 4387–4392 doi:10.1073/pnas.0408648102.

53. Collins SL, Knapp AK, Briggs JM, Blair JM, Steinauer EM (1998) Modulation of diversity by grazing and mowing in native tallgrass prairie. Science 280: 745–747. doi: 10.1126/science.280.5364.745

There are several supplemental files that are not available in this version of the article. To view this additional information, please use the citation on the first page of this chapter.

CHAPTER 6

Seasonal Energy Storage Using Bioenergy Production from Abandoned Croplands

J. ELLIOTT CAMPBELL, DAVID B. LOBELL, ROBERT C. GENOVA, ANDREW ZUMKEHR, AND CHRISTOPHER B. FIELD

6.1 INTRODUCTION

The production of electricity from biomass could provide a critical back-up energy source to renewable energy portfolios that are currently dominated by intermittent wind and solar energy resources [1]. The stored chemical energy in biomass can be deployed to produce electricity on demand, reducing the overall intermittency of a renewable energy portfolio. The storage capacity of biomass is particularly important for the seasonal intermittency of wind and solar energy which are not easily addressed by alternative storage schemes such as pumped hydropower, thermal energy storage, compressed air energy storage, flow batteries, fuel cells, flywheels, or superconducting magnetic energy storage [2, 3]. Seasonal variation in wind and solar energy production require much larger amounts of energy

Seasonal Energy Storage Using Bioenergy Production from Abandoned Croplands. © *Campbell JE, Lobell DB, Genova RC, Zumkehr A, and Field CB.* Environmental Research Letters *8,3 (2013), doi:10.1088/1748-9326/8/3/035012. Licensed under Creative Commons Attribution 3.0 Unported License, http://creativecommons.org/licenses/by/3.0/.*

storage than required to address short-term intermittency in wind and solar energy production as well as energy storage over longer periods.

In addition to energy storage, biomass electricity has the unique potential of providing a carbon-negative and highly efficient approach to bioenergy production. Mandated reductions in carbon emissions may require technological solutions that cannot be met by carbon-neutral solutions alone. Biomass electricity can provide a carbon-negative solution when biomass cultivation and energy conversion technologies are coupled to carbon capture and sequestration technologies [4, 5]. Alternatively, there are approaches to bioenergy production that can results in large life-cycle emissions of greenhouse gases (GHG) due to fossil fuel energy consumed during the cultivation of biomass, fossil fuel energy consumed during the conversion of biomass into useful energy forms, and land-use change [6, 7]. Creating a carbon-negative approach to biomass electricity will require careful consideration of these cultivation, energy conversion, and land-use factors. Biomass electricity has also been shown to provide a highly efficient approach to providing renewable transportation energy that is complementary to existing liquid fuel approaches [8–11].

Despite the important role that bioenergy may play in future energy systems, the economic and environmental outcomes of bioenergy remain highly uncertain. This uncertainty is dominated by bioenergy land-use effects that may disrupt ecological systems and food economies but may also contribute to rural development and provide new options for enhancing soil quality, water resources, and biodiversity [4, 12–15]. Field studies suggest a path to reduce negative land-use impacts through the application of abandoned agriculture lands for biomass cultivation as opposed to prime agriculture lands [4]. A global-scale analysis of abandoned lands estimated the bioenergy potential of these lands as 32–41 EJ or 7%–8% of primary energy demand [16, 17]. While these coarse global estimates are useful for large-scale planning, more practical technological and policy applications will require regional, high-resolution information on biomass availability.

Here we present county-level estimates of the magnitude and distribution of abandoned agriculture lands in the US using remote sensing, agriculture inventories, and land-use modeling. Furthermore, we explore one potential application of these data to quantifying the seasonal energy storage that could be provided by bioenergy to compensate for the seasonal intermittency of wind and solar power. The county-level land-use data resulting from this study are extended to a gridded product in a companion paper [18]. The land-use data resulting from this analysis are available online (https://eng.ucmerced.edu/campbell/).

6.2 METHODS FOR LAND-USE AVAILABILITY AND BIOENERGY POTENTIAL

We employ a GIS-based modeling approach to develop maps of land availability, biomass yields, and bioenergy production with a county-level resolution for the US domain. The GIS model was developed using the ArcGIS spatial analyst extension to conduct the raster and vector calculations described below. The land-use model quantifies the spatial distribution of abandoned agriculture. Abandoned agriculture is divided into abandoned cropland and abandoned pasture. Here available abandoned agriculture lands are defined as an area of land that was once classified as cropland or pasture and is currently not classified as cropland, pasture, forestland, or urban areas. The input data for the land-use analysis is the USDA county-level cropland database which includes cropland areas for each county from the years 1850 to 1997 [19]. Between 1850 and 1940, the data are in 10-year increments. Between 1940 and 1997 the data are in 4 or 5 year increments. The abandoned cropland area for each county is the difference between maximum area over the time series for each county (years 1850–1997) and the area in the final time step (year 1997),

$$A_i = \max_{1850 \le t \le 1997}(C_{i,t}) - C_{i,1997} \tag{1}$$

where A_i is the abandoned cropland areas, C is the cropland area, and i and t are the indices for county and year, respectively. Note that in equation (1), the maximum cropland areas from each county are not necessarily from the same year but may be from a range of years. The county-level cropland data had erroneous spikes [19] for less than 5% of the data which were removed prior to the analysis. Estimates of abandoned pasture land as well as estimates of areas of land that was converted from agriculture to urban or forestlands were obtained from a previous analysis [16, 17]. More recent cropland area changes after 1997 were explored using the USDA/ NASS database and found to be small relative to area changes from 1850 to 1997. However, recent work presents a small but rapid expansion of cropland areas from 2006 to 2011 which suggests that our analysis may overestimate abandoned croplands in some localities [20].

The county-level data suffered from a change in land-use definitions between 1940 and 1945 which introduces an artificial decline in cropland area [19, 21]. The period of 1945–1997 has a more restrictive definition of cropland used for pasture than for the period of 1850–1940. We adjusted for the definition change by subtracting the 1940–1945 area change from the county areas for years 1850–1940.

There are several sources of uncertainty in these land availability estimates. The spatial resolution of the croplands (county-level) is not consistent with the pasture, forest, and urban land (5 min × 5 min) data. Furthermore the crop and pasture data have a different format than the forest and urban area data. The crop and pasture data are density data, providing the per cent of each pixel that is occupied by crops or pastures. Alternatively, the forest and urban data classify each pixel as entirely forest or urban or other land cover without providing the fraction of the pixel that is covered by such land cover. Validation data are not currently available for determining the magnitude of this uncertainty associated with the range of spatial resolutions and classification schemes. However, we also report abandoned cropland areas that are not combined with pasture areas and are not filtered using forest and urban areas in order to provide an upper estimate for land availability that is not associated with these uncertainties.

We compare these new abandoned cropland area estimates to previous estimates that were based on the relatively coarse data from the SAGE and HYDE global gridded databases [22, 23]. HYDE crop and pasture estimates range from 1700 to 2000 in 10-year increments. We developed HYDE-based abandoned cropland areas using only data from 1850 to 2000. Estimating abandoned cropland from HYDE using data from 1850 to 2000 yielded the same results as using data from 1700 to 2000. SAGE crop areas range from years 1700 to 1992 in 10-year increments. Abandoned crop estimates were calculated using data from 1850 to 1992.

Biomass yields and energy conversion efficiencies are based on approaches applied in previous global assessments [16, 17]. Crop yields are based on the CASA primary production model of natural vegetation and are in the range of observed yields for the candidate biomass crop switchgrass [24] but lower than the relatively sparse observations available for the candidate crop Miscanthus [25]. We used the CASA model to spatially extrapolate reported biomass yields for switchgrass and Miscanthus [25]. The ratios of the observed Miscanthus and switchgrass yields in Illinois with respect to the CASA yield simulations in Illinois were applied to the CASA yield map to provide a map of Miscanthus and switchgrass yields. The above-ground yields at the Illinois site were 29.6 t dry biomass ha^{-1} yr^{-1}, 10.4 t dry biomass ha^{-1} y^{-1}, and 9.9 t dry biomass ha^{-1} y^{-1} for the Miscanthus plot, switchgrass plot, and CASA grid cell, respectively. These yield estimates may provide an upper estimate of biomass yields due to the potential for soil degradation on abandoned lands.

6.3 SEASONAL STORAGE AND DEMAND MODEL

Recent work by Converse considers a range of energy storage options for compensating for the seasonal-scale intermittency of wind and solar energy production [2]. This analysis considers the seasonal electricity storage requirements for a renewable electricity system that uses no fossil fuels. The seasonal energy storage requirement was based on current national wind and solar production data for the US domain. This approach is designed to assess the seasonal energy storage requirement for a future US energy scenario in which wind and solar energy are the dominant energy sources. The annual wind and solar production is assumed to be equal to the annual energy demand. Three renewable energy scenarios are examined including only wind production, only solar production, and the energy production divided evenly between wind and solar. This approach assumes that current seasonal variations of electricity demand, solar production, and wind production in the future will be similar to the current seasonal variation rather than projecting alternative seasonal capacity factors. This approach only estimates energy storage requirements at the seasonal scale and does not consider storage requirements at different timescales (e.g., diurnal storage requirements) or the potential for existing hydropower to provide storage or back-up power. Alternatively, if a larger generation capacity had been assumed then the seasonal storage requirements would be reduced.

We quantify the seasonal energy storage requirement using DOE Energy Information Administration (EIA) monthly data for current electricity demand, current wind production, and current solar production at the level of three regional power grids (east grid, Texas/ERCOT grid, and west grid) [26]. Solar energy is not analyzed for the ERCOT grid due to the relatively small scale and inconsistency of EIA data for this region. The simulated monthly storage is,

$$S_{m+1} = S_m - (D_m - P_m) \tag{2}$$

where S is the monthly storage (normalized to annual demand), D is the electricity demand (normalized to annual demand), P is the electricity production (normalized to annual production), and m represents the month. The annual seasonal storage requirement (SD) is,

$$SD = \max_{1 \le m \le 12}(S_m) - \min_{1 \le m \le 12}(S_m) \tag{3}$$

The annual storage requirement is the storage capacity that would be required to offset energy deficits due to seasonal intermittency. This simple model provides an estimate of the annual storage capacity needed for storage technologies that can store excess wind and solar energy during months when wind and solar production exceed demand. When bioenergy is the back-up energy technology, the surplus wind and solar energy cannot be stored and the annual energy storage is simply the sum of the monthly deficits during months when production is less than demand.

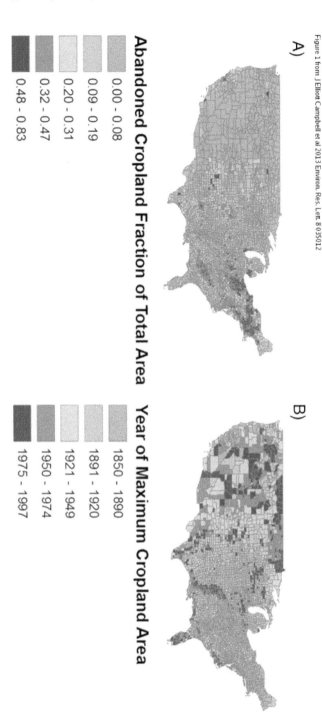

Figure 1 from J Elliott Campbell et al 2013 Environ. Res. Lett. 8 035012

FIGURE 1: Abandoned cropland (A) and year of maximum cropland area (B). Abandoned cropland areas are presented as the fraction of county area and are total abandoned areas not excluding conversion to forest or urban areas.

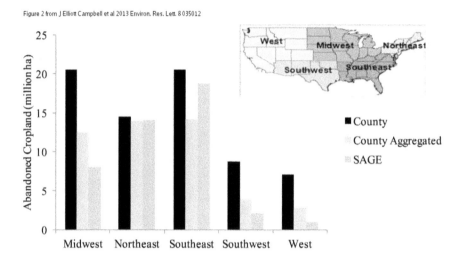

Figure 2 from J Elliott Campbell et al 2013 Environ. Res. Lett. 8 035012

FIGURE 2: Regional abandoned cropland areas based on county-level data, county-level data aggregated to the state-level, and previous global study based on state-level data (SAGE).

6.4 LAND AVAILABILITY RESULTS

Abandoned cropland area (not adding abandoned pasture or removing croplands converted to forest or urban areas) for the years 1850–1997 is found to be 94.5 Mha (million hectares) using the uncorrected county-level USDA cropland data. However 12 of the counties have anomalous cropland areas in which a spike appears in the timeline that is as much as seven times the total land area in a county. Removing these anomalous spikes results in 99.94% of counties having cropland areas that are less than the total county area and all counties having cropland areas that are less than 105% of total county area. The abandoned cropland estimate resulting from this revised data set is 90.9 Mha. Adjusting for the land-use definition change (1940–1945) results in an abandoned cropland area of 71 Mha or about 41% of the current cropland area (figure 1). There is only a 2% difference in the estimated area if the 1940–1945 data gap is filled using one time step before and after the gap or two time steps before and after the gap.

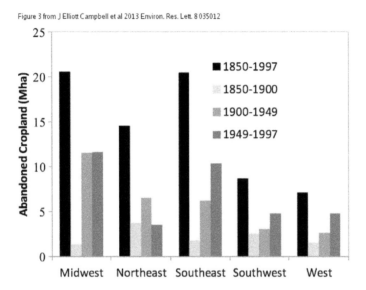

Figure 3 from J Elliott Campbell et al 2013 Environ. Res. Lett. 8 035012

FIGURE 3: Abandoned cropland areas based on temporal subsets of the data spanning year 1850–1900, 1900–1949, and 1949–1997.

Our abandoned cropland area estimate of 71 Mha is larger than previous estimates that were based on coarse global data sets (figure 2). We used a similar land-use modeling approach with the SAGE global gridded databases which result in 44 Mha of abandoned cropland. These global gridded databases were derived from state-level rather than county-level census data. Our county-level results may be larger than previous state-level results because of aggregation effects in the state-level results. To quantify this source of error we aggregated the county-level data to the state-level and repeated the analysis which resulted in a greater similarity to the SAGE areas (figure 2). The aggregation effect is small in the northeast relative to other regions of the US, perhaps due to the smaller size of the states in this region. One possible driver of this aggregation effect is that counties with cropland abandonment may be masked by counties with cropland area growth when both exist within the same state. If aggregation effects are also occurring at the county-level then our county-level results may also underestimate the magnitude of the abandoned areas.

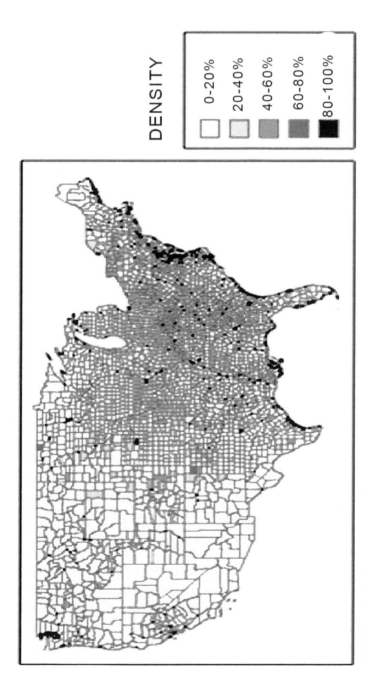

FIGURE 4: Total abandoned agriculture as a per cent of county area. Abandoned agriculture includes abandoned cropland and abandoned pasture but excludes abandoned lands that have been converted to forests or urban areas.

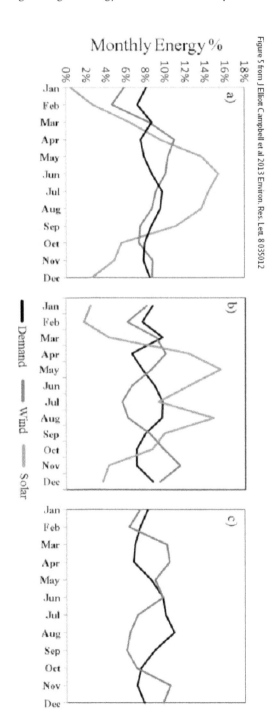

FIGURE 5: Monthly wind electricity production, solar electricity production and electricity demand as a percentage of the annual totals for each. Seasonal energy profiles for states within the (a) Western electric grid, (b) Eastern electric grid and (c) ERCOT (Texas) electric grid.

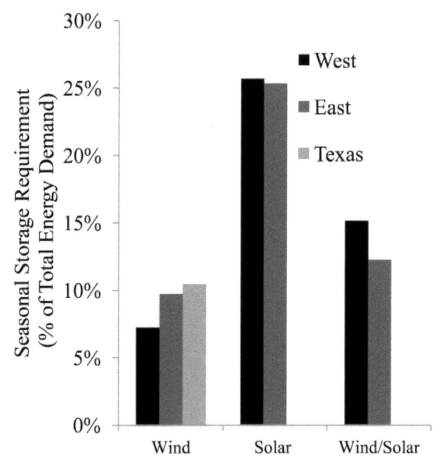

FIGURE 6: Seasonal electricity storage required for Western electric grid, Eastern electric grid and ERCOT (Texas) electric grid based on an energy production system that is entirely wind, entirely solar, or a 50% mix of wind and solar.

The timing of cropland abandonment is related to the year of maximum crop area which is on average 1933 ($\pm$38) for the US domain (figure 1(B)). Restricting the land-use analysis to the years 1900–1997 results in 93% of the abandoned area estimate as opposed to using the entire data set for years 1850–1997 (figure 3).

The abandoned agriculture area accounting for cropland abandonment, pasture abandonment, and excluding conversion of these lands to forest or urban areas are plotted in figure 4. The total abandoned agriculture is 99 Mha, which is considerably larger than the US results using global data of 51–67 Mha [16].

6.5 SEASONAL ENERGY STORAGE RESULTS

The monthly time series for energy demand, wind energy, and solar energy are plotted in figure 5. The seasonal amplitude of solar energy is considerably larger than the seasonal amplitude of wind energy and energy demand. While the amplitude of the wind energy is relatively low, the wind variation is not in phase with the demand variation.

The required seasonal storage requirement is plotted as a fraction of the total electricity demand in figure 6. The required storage varies from 7% to 26% of annual electricity demand suggesting that seasonal storage would be an important component of a renewable energy system based on wind and solar. Because of the relatively low seasonal variability of wind energy, it follows that required seasonal storage by the wind energy scenario is low relative to the solar scenario while the wind and solar mixed scenario falls in between. Our results have a similar overall range as previous results based on national-scale data [2], but differ in terms of the storage required for each specific pathway and add a regional component to the analysis.

The capacity of biomass to meet the seasonal storage requirement is plotted for a range of biomass feedstocks, energy production scenarios, and regions in figure 7. In the estimates shown in figure 7, the biomass energy values represent biomass energy that can be potentially obtained from available abandoned cropland (not including abandoned pasture) within the respective region. Bioenergy can meet more than half of the storage requirements for most cases considered if all the available biomass grown on abandoned croplands is utilized. Bioenergy can meet all of the seasonal storage requirements for the 100% solar production scenario only for the most optimistic assumption regarding biomass crop yields.

6.6 DISCUSSION AND CONCLUSIONS

Previous work suggests the potential for US abandoned agriculture lands to provide a sustainable land resource for bioenergy production [4, 16]. Here we find that high-resolution land-use databases provide a 61% larger area estimate of abandoned croplands than previously reported. Our larger area estimate is due to the use of more spatially resolved input data relative to the state-level data applied in previous global studies. Furthermore, these regional data are better suited for technical bioenergy studies and policy investigations as opposed to the spatially coarse data from previous work.

These results may also suggest a larger uncertainty in abandoned agriculture than indicated by previous work. Previous studies have used two alternative land-use databases, SAGE and HYDE, to estimate abandoned agriculture lands [16, 17]. The abandoned cropland estimates for the US for the SAGE and HYDE analysis were 69 Mha and 44 Mha, respectively. The results of our county-level work are 91 Mha suggesting a larger range for the uncertainty of abandoned cropland estimates than what is expected from considering SAGE and HYDE data alone.

Further analysis of the bioenergy potential should consider available land resources, yield estimates, and energy conversion efficiencies. While our analysis is focused on abandoned croplands as a sustainable domain for bioenergy, other studies have argued that additional land resources may be sustainably utilized including marginal croplands, abandoned pasture lands, forests, and waste biomass [27, 28]. Factors including biomass harvest logistics, biomass transport, biomass storage, and degradation of soil and water resources over time would tend to result in smaller estimates of available bioenergy than the estimates presented here. Finally it has been argued that alternative uses of available lands should be considered including additional wind and solar production [29].

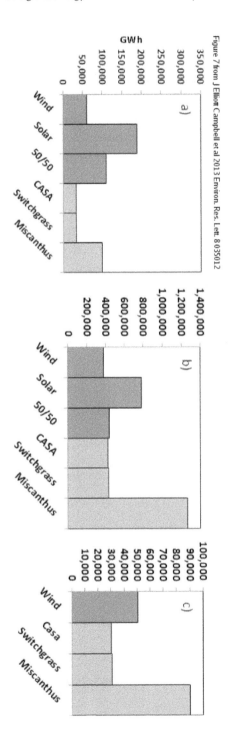

Figure 7 from J Elliott Campbell et al 2013 Environ. Res. Lett. 8 035012

FIGURE 7: Annual electricity energy storage deficit and bioenergy availability for a range of biomass feedstock crops. Storage needs by resource are in red (if 100% of the annual demand was met only by the specified resource) and estimates of available energy from CASA NPP estimates, Switchgrass and Miscanthus at a 30% conversion efficiency from biomass to electricity in green. Energy storage is plotted for (a) Western electric grid, (b) Eastern electric grid and (c) ERCOT (Texas) electric grid.

We applied our new land-use data to consider the role of bioenergy in providing seasonal energy storage. Seasonal energy storage is required to address the intermittency of a future energy production system that may be based on wind and solar energy without the use of fossil fuel energy. Examining seasonal storage requirements for a hypothetical future energy system may be useful for informing the development of technologies and infrastructure investments that are needed to bridge the near-term energy system to the endpoint energy system. Our simple approach suggests the need for a large seasonal storage capacity (7%–26% of energy demand). We found that bioenergy could provide most of the seasonal storage requirements for most energy pathways, though a system dominated by solar energy requires relatively optimistic assumptions regarding biomass yields to satisfy the storage requirement.

REFERENCES

1. Campbell J E 2010 Integrating the sustainable energy portfolio: a bioenergy perspective Carbon Manag. 1 15–7
2. Converse A O 2012 Seasonal energy storage in a renewable energy system Proc. IEEE 100 401–9
3. Ibrahim H, Ilinca A and Perron J 2008 Energy storage systems—characteristics and comparisons Renew. Sustain. Energy Rev. 12 1221–50
4. Tilman D, Hill J and Lehman C 2006 Carbon-negative biofuels from low-input high-diversity grassland biomass Science 314 1598–600
5. Wise M et al 2009 Implications of limiting CO_2 concentrations for land use and energy Science 324 1183–6
6. Farrell A E et al 2006 Ethanol can contribute to energy and environmental goals Science 311 506–8
7. Searchinger T et al 2008 Use of US croplands for biofuels increases greenhouse gases through emissions from land-use change Science 319 1238–40
8. Campbell J E and Block E 2010 Land-use and alternative bioenergy pathways for waste biomass Environ. Sci. Technol. 44 8665–9
9. Campbell J E, Lobell D B and Field C B 2009 Greater transportation energy and GHG offsets from bioelectricity than ethanol Science 324 1055–7
10. Hedegaard K, Thyø K A and Wenzel H 2008 Life cycle assessment of an advanced bioethanol technology in the perspective of constrained biomass availability Environ. Sci. Technol. 42 7992–9
11. Clarens A F et al 2010 Environmental life cycle comparison of algae to other bioenergy feedstocks Environ. Sci. Technol. 44 1813–9

12. Parish E S et al 2012 Multimetric spatial optimization of switchgrass plantings across a watershed Biofuels, Bioprod. Biorefin. 6 58–72

13. Gopalakrishnan G, Cristina Negri M and Salas W 2012 Modeling biogeochemical impacts of bioenergy buffers with perennial grasses for a row-crop field in Illinois GCB Bioenergy 4 739–50

14. Gopalakrishnan G, Negri M C and Synder S W 2011 Redesigning agricultural landscapes for sustainability using bioenergy crops: quantifying the tradeoffs between agriculture, energy and the environment Biomass and Energy Crops IV (Aspects of Applied Biology 112) (Warwick: Association of Applied Biologists) pp 139–46

15. Fargione J et al 2008 Land clearing and the biofuel carbon debt Science 219 1235–8

16. Campbell J E et al 2008 The global potential of bioenergy on abandoned agriculture lands Environ. Sci. Technol. 42 5791–4

17. Field C B, Campbell J E and Lobell D B 2008 Biomass energy: the scale of the potential resource Trends Ecol. Evol. 23 65–72

18. Zumkehr A and Campbell J E 2013 Historical US Cropland areas and the potential for bioenergy production on abandoned Croplands Environ. Sci. Technol. 47 3840–7

19. Waisanen P J and Bliss N B 2002 Changes in population and agricultural land in conterminous United States counties, 1790 to 1997 Glob. Biogeochem. Cycles 16 1137

20. Wright C K and Wimberly M C 2013 Recent land use change in the Western Corn Belt threatens grasslands and wetlands Proc. Natl Acad. Sci. 110 4134–9

21. Ramankutty N, Heller E and Rhemtulla J 2010 Prevailing myths about agricultural abandonment and forest regrowth in the United States Ann. Assoc. Am. Geogr. 100 502–12

22. Ramankutty N and Foley J A 1999 Estimating historical changes in global land cover: croplands from 1700 to 1992 Global Biogeochem. Cycles 13 997–1027

23. Goldewijk K K and Ramankutty N 2004 Land cover change over the last three centuries due to human activities: the availability of new global data sets GeoJournal 61 335–44

24. Schmer M R et al 2008 Net energy of cellulosic ethanol from switchgrass Proc. Natl Acad. Sci. USA 105 464–9

25. Heaton E A, Dohleman F G and Long S P 2008 Meeting US biofuel goals with less land: the potential of Miscanthus Glob. Change Biol. 14 2000–14

26. DOE 2012 Electricity (Washington, DC: US Energy Information Administration) (www.eia.gov/electricity/)

27. Cai X, Wang D and Zhang X 2011 Land availability for biofuel production and the environmental impacts Environ. Sci. Technol. 45 334–9

28. DOE 2005 Biomass as Feedstock for a Bioenergy and Bioproducts Industry: The Technical Feasibility of a Billion-Ton Annual Supply (Oak Ridge, TN: Oak Ridge National Laboratory)

29. Geyer R, Stoms D and Kallaos J 2013 Spatially-explicit life cycle assessment of sun-to-wheels transportation pathways in the US Environ. Sci. Technol. 47 1170–6

PART II

SECOND-GENERATION BIOFUELS AND SUSTAINABILITY

CHAPTER 7

Biodiesel from Grease Interceptor to Gas Tank

ALYSE MARY E. RAGAUSKAS, YUNQIAO PU,
AND ART J. RAGAUSKAS

7.1 INTRODUCTION

Over the last decade the sustainable conversion of bioresources to biofuels has become a global pursuit that is being tailored to regional resources and local needs [1, 2]. Today's biofuel successes are often contingent on using abundant and productive starch- and sucrose-based crops that has been challenged on "food-or-fuel" concerns which will only increase in the future as the global population grows [3, 4]. Key commercial breakthroughs in replacing significant amounts of petroleum-based fuels with renewable, nonfood bioresources, will come from translational research directed at reducing the recalcitrance of lignocellulosics for sustainable 2nd and 3rd generation biofuels [5] and overcoming the barriers to algae-derived biofuels [6] with positive life-cycle assessment (LCA) performance parameters. While these efforts are continuing to accelerate with technical issues being addressed, smaller successes also continue to evolve. For example,

Biodiesel from Grease Interceptor to Gas Tank. © *Ragauskas AME, Pu Y, and Ragauskas AJ.* Energy Science and Engineering **1**,1 (2013), DOI: 10.1002/ese3.4. *Licensed under a Creative Commons Attribution 3.0 Unported License, http://creativecommons.org/licenses/by/3.0/.*

the recovery of yellow grease from commercial cooking facilities is a well-documented success in which spent fryer oil, containing a mixture of plant and animal fats, is stored, collected, and then transported to a central process site in which it is purified and transesterified to biodiesel (see Fig. 1) [7].

After purification, FAME is sold as biodiesel most often as B2, B5, or B10 fuel blends. The excess alcohol in FAME generation is recovered and recycled, whereas the glycerin is recovered and sold into established commercial markets. Given the volume of glycerin generated, a growing research effort is now being directed at using this chemical as a feedstock for alternative chemicals and polymers [8].

In many countries, companies will not only pay for used restaurant cooking oils but also from food processing plants, animal processing centers, supermarkets—almost any business that produces cooking oil waste or used fryer oil. It is now indeed difficult to envisage that these fats, oils, and greases (FOG) were once considered a waste product. Key to this conversion is a low level of free fatty acids (FFA) which are usually cited to be 15% or lower in yellow grease. The presence of FFA complicates the transesterification reaction as this is most often accomplished using an alkaline system (i.e., frequently $NaOCH_3$) and FFA neutralize the base and lead to soap formation. A commonly cited practice with yellow grease with a moderate FFA content is to blend this feedstock with fats thereby lowering the overall FFA content to 1–3%. Literature reports indicate that a FFA content of <2.5% does not yield significant processing difficulties [9], although Mittlebach reported the level of FFA should be no more than ~1% for the alkaline-catalyzed transesterification reactions [10]. Alternative approaches to controlling FFA content include steam stripping [11], caustic stripping, solvent extraction [12], glycerolysis, or acid esterification, with the latter approach being highlighted by several biodiesel companies [13-17].

As the demand for biodiesel increases and viable yellow grease sources have been secured, entrepreneurs have begun to pursue the conversion of select brown grease resources (i.e., FFA content >15%) to biodiesel which is the focus of this review. The attractive features of commercially recovered brown grease are fourfold [18]:

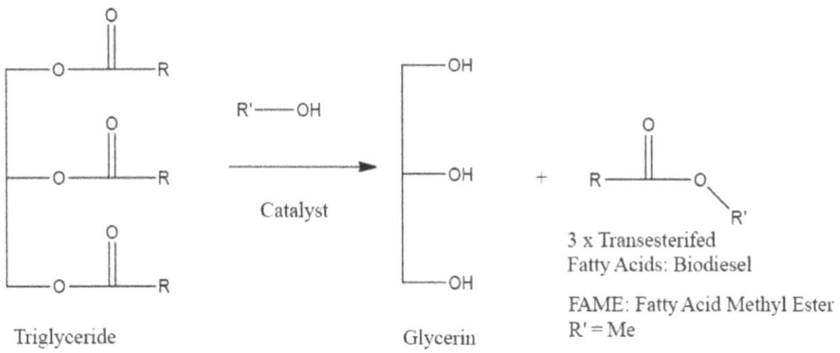

FIGURE 1: Biodiesel transesterification reaction.

- Lower feedstock cost.
- Large volume of resource available.
- Governmental mandates requiring collection and processing of select brown greases.
- Avoidance of "food versus fuel" concerns while contributing to the development of renewable fuels.

7.2 FOG COLLECTION

One of the biggest sources of brown grease is the material trapped and recovered in grease inceptors/traps that many commercial food processing centers are mandated to have. Grease abatement plumbing devices are usually nonmechanical gravity separation flow-through devices that facilitate the recovery of grease and food solids from aqueous waste streams. Depending on the size of food processing operations, modern building/business codes often require the installation of grease traps (~50 L) or interceptors (~3780–7570 L) [19, 20]. Grease interceptors are multicompartment chamber devices where the aqueous grease containing flow is retained long enough so that grease and some solids can rise to the water surface and most of the solids settle to the bottom. The clarified water is then eventually discharged to a sanitary sewer system (see Fig. 2).

Given their size, grease interceptors are usually located below ground and outside the food preparation area. The American Society of Mechanical Engineers (ASME) standard A112.14.3 requires that grease interceptors remove a minimum of 90% of the incoming FOG [21]. Grease traps are designed to retain a small amount of grease, usually servicing from 1 to 4 plumbing fixtures. A recent study has suggested that FOG removal may be less than these target values, as a pilot system suggested that retention-based grease interceptors achieved ~80% FOG removal and flow-based grease interceptors removed <50% FOG [22]. A review of the literature indicates that grease interceptors should be pumped out by permitted transporters, at a minimum every 90 days, whereas shorter maintenance cycles (i.e., 30 or 60 days) have been reported depending on facility's service volume [23]. Alternatively, some municipalities have implemented a "25% rule" that requires the grease abatement unit be emptied when they reach 25% of the design capacity [24].

The need for active management and control of trap FOG is due to the detrimental impact on sewer systems. The U.S.A. EPA has estimated that 23,000 to 75,000 sanitary sewer overflows (SSO) per year occurred in the United States in the 2001–2003 time period of which 48% of the SSOs were caused by blockages and 47% of these blockages were grease related, leading to ~5000 to 17,000 FOG-caused SSOs nationwide annually [25-27]. Furthermore, it has been proposed that FOG contributes to sewer plant blockages and pump station failures [28]. These overall concerns for trap FOG in sewer systems have also been reported for a variety of other major metropolitan cities globally. For example:

- City of Dublin "Drainage maintenance records indicate that FOG is a serious problem in areas where there are concentrations of Food Service Establishments (FSE) such as pubs, restaurants, hotels, takeaways, convenience stores, etc." [29].
- 2012, Metro Vancouver approved a bylaw that mandates new requirements for grease interceptors or grease traps [30].
- To reduce sewer blockages, South East Water in Australia implemented a three-phase grease reduction program which involved the use of grease interceptor and mandated pump-outs [31].

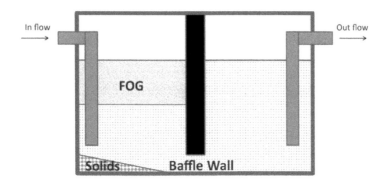

FIGURE 2: Simplified illustration of fats, oils, and greases abatement device.

The grease trap waste has been reported to vary substantially depending on the source but a broad description of these abatement collection systems would include three phases: a top floatable layer rich in FOG; a middle aqueous layer that is organic rich; and bottom sludge containing food particles and other settleable solids. The trap material is about 95% water, 3% solids, and 2% FOG. Sato et al. surveyed 27 different restaurant grease samples and reported that the average chemical oxygen demand (COD) levels for the floatable layer was 478,000 mg/L, 66,200 mg/L for the middle layer and 1,007,061 mg/L for the sludge phase [32]. Historically, this material was collected and landfilled, although other options included land application, compositing, rendering for lubricants/soaps, or incineration. Direct disposal is becoming more challenging due to legislative regulations and overall decreased access to inexpensive landfill options. The two most attractive future applications for grease trap waste are anaerobic codigestion of FOG to methane/biogas or biodiesel production. The former pathway was extensively reviewed by Long et al. [33] in 2012 and this review examines opportunities in the biodiesel field. The latter route could be especially attractive to those regions of the globe where natural gas prices are low and biodiesel prices are high. In addition, for select developing nations, it was been reported that the market demand for FOG is very low and despite legislative efforts this material (i.e., gutter oil) has been reintroduced into the food services industry [34]. Clearly, there is a need to develop practical market-driven solutions that will provide viable market-driven outlets for trap FOG which the biodiesel route could readily provide.

7.3 FOG FROM GREASE TRAPS/INTERCEPTORS

A recent National Renewable Energy Laboratory report estimated that in the United States, FOG is generated at a rate of 7.1 L FOG/person/year [35]. But Long et al. [33] has argued this may overstate the recoverable amounts of FOG as this value includes FOG in the sewer system hence the amount of annually recoverable FOG generated in the United States should be ~2.2 billion L/year. On a more regionally basis, literature reports indicate that substantial amounts of trap FOG are generated in most major metropolitan cities as highlighted in Table 1 [36].

TABLE 1: Select U.S.A. urban trap grease generation amounts [37]

City/Sate	Annual trap grease generation (m. tons)
Sacramento/CA	7,530
Denver/CO	7,200
Boston/MA	15,200
Washington/DC	22,700
Memphis/TN	8,400
Macon/GA	2,700

Given the fact this material must be collected, is relatively low cost, and present in large volumes in major metropolitan regions, it opens the opportunity for trap FOG to biodiesel processing centers. The chemical components that contribute to grease trap/interceptor FOG is highly variable, as to be expected, given the numerous sources that can contribute to this bioresource. Nonetheless, there are several published articles that highlight the components present in this resource as summarized in Table 2.

TABLE 2: Reported fatty acids present in trap FOG

Compound	Average of 27 restaurant grease samples % [32]	Grease trap influent for Korea O Barbecue restaurant % [38]	Cafeteria and restaurant, Thailand % [39]	Polish meat processing plant % [40]	Canteen National Singapore University % [41]	Trap grease provided by Guangzhou E.P.A. of China [42]
Caprylic Acid – C8:0	0.9		<0.01			
Capric Acid – C9:0		0.04				
Decanoic Acid – C10:0	0.9	0.1	<0.01			
Undecylic Acid – C11:0		ND				
Lauric Acid – C12:0	1.3	0.58	0.03			
Myristic Acid – C14:0	8.4	0.56	0.02	0.01	1.3	1.16
Pentadecylic – C15:0		0.02				
Palmitic Acid – C16:0	23.1		25.8	28.83	38.3	30.38
Palmitoleic Acid – C16:1			0.03		1.2	1.42
Margaric Acid – C17:0		1.06				
Stearic Acid – C18:0	9.8	4.71	0.06	16.31	7.2	6.02
Nonadecylic Acid – C19:0		ND				
Arachidic Acid – C20:0	2.1	0.17				
Palmitoleic Acid – C16:1ω7c		0.73				
Cis-9-heptadecenoic Acid – C17:1ω8c		3.12				
Linoleic Acid – C18:2	15.3		19.9		15.1	18.83
Alpha linoleic acid – C18:2ω6,9c		20.92				

TABLE 2: *Cont.*

Compound	Average of 27 restaurant grease samples % [32]	Grease trap influent for Korea O Barbecue restaurant % [38]	Cafeteria and restaurant, Thailand % [39]	Polish meat processing plant % [40]	Canteen National Singapore University % [41]	Trap grease provided by Guangzhou E.P.A. of China [42]
Gamma linoleic acid – C18:3ω6,9,12c		3.29				
Linolenic Acid – C18:3	–					1.31
Palmitoleic Acid – C16:1	–		<0.01			
Oleic Acid – C18:1ω9c	36.1	22.87	39.6	53.37	36.9	38.39
Cis-13-nonadeconic Acid – C19:1ω6c		2.56				
Gondoic Acid – C20: 1ω9c		0.28				
Arachidic Acid – C20:0			<0.01	<0.01		
Behenic Acid – C22:0			<0.01			
Lignoceric Acid: C24:0			<0.01			
60–80 minor free fatty acids		31.40				2.49

As summarized in Table 2 and other reports [43], often the predominant saturated fat is palmitic, primary unsaturated fat is oleic, and the major polyunsaturated fat is linoleic acid. A report by Robbins et al. [44] indicates that the free fatty acid content of trap greases can vary from 35% to 100%. Other reports have provided lower FFA values such 26.2% from trap grease FOG isolated from restaurants and cafeterias in Thailand [39]. Although the trap FOG collected in grease abatement plumbing devices may have started as a yellow grease, the high and varied levels of FFA is not surprising given the assorted cuisines, collection systems/operating environments, time of collection, comingled containments, presence of food particles, salts, alkaline cleaning products, detergents, and biological agents. For example, the presence of detergents and sanitizers has been reported to enhance the hydrolysis of triglycerides yielding FFA [20].

TABLE 3: Volatile organic compounds found in trap grease [45]

	Reported range for 17 samples (μg/L)
Acetone	157.0–713.0
Benzene	3.8–10.5
2-Butane	92.6–1040.0
Carbon disulfide	1.6–172.0
Chloroethane	11.2–62.0
Chloroform	2.9–655.0
Chloromethane	2.6–27.0
1,2-Dichloroethene	28.2–102.0
Ethylbenzene	11.5–98.8
2-Hexanone	12.0
Methylene chloride	2.0–288.0
Styrene	4.9–17.3
Tetrachloroethene	31.5–8510.0
Toluene	13.9–1370
Trichloroethene	146.0–600.0
Xylene	16.6–687.0

Given the variability of potential nonfood inputs into grease traps (i.e., cleaning solutions, municipal water trace containments, pesticides, etc.), it is well appreciated that this resource could contain a variety of other materials and the open literature does provide some guidance to these elements as summarized in Tables 3, 4.

The reported chlorinated compounds were attributed to reactions of dissolved organics in water chlorination programs and/or biological routes.

To better understand how these containments need to be handled, the following section will examine the commonly followed procedure for the conversion of trap FOG to biodiesel, followed by a brief description of the purification steps employed.

TABLE 4: Inorganic elements detected in trap grease [45]

	Reported range for samples tested[a]
Calcium (mg/L)	240–37,359
Copper (mg/L)	1.8–1092
Iron (mg/L)	70–577
Lead (mg/L)	1.1–81.7
Magnesium (mg/L)	30–3575
Phosphorous (mg/L)	6.5–326
Potassium (mg/L)	14–11,968
Sodium (mg/L)	34–54,348
Zinc (mg/L)	21–467
Nitrogen (mg/L)	0.2–6.3
Chloride (mg/L)	0–4275
% Sulfur	0.1–0.3

[a]*In some cases, values varied depending on testing laboratory, in which case the highest and lowest values are reported; No heavy metals found in rendered lipids.*

7.4 TRAP GREASE FOG TO BIODIESEL

The chemistry of converting fatty acid triglycerides to biodiesel is well established from a commercial perspective and remains actively pursued as a research topic. Industrially, the most commonly practiced route is via a base-catalyzed transesterification methodology using methanol and sodium or potassium hydroxide due to the low cost and ease of processing [46, 47]. As discussed earlier, this approach requires a low level of FFA (i.e., <2%) as soap formation leads to a wasting of base and numerous process complications.

TABLE 5: Acid-catalyzed esterification of free fatty acids in trap grease FOG

Methanol:Oil v/v	H_2SO_4 % v/v	Reaction Time/h	Reaction Temp./°C	% Residual acid content (mg KOH/g)
Range studied				
0.34–0.51	1.00–5.02	2.0–7.4	65	20.3–2.2
Preferred conditions				
0.42	2.50	4.0–7.4	65	<2.6

Note: Starting FOG acid value: 52.13 KOH/g; water removed by distillation) as reported by Karnasuta et al. [39].

7.5 CONVERSION OF FOG TO FAME

For trap FOG with free fatty acid contents >20%, the usually process "trick" used with yellow grease to lower the content of FFA (i.e., adding a high quality fat) is not operationally possible. To address this challenge, the most commonly applied conversion technology is to treat this material with an acidic methanol solution which converts the FFA to methyl esters followed by a conventional base-catalyzed transesterification step (see Fig. 3).

Several recent publications provide guidance as to optimal reaction conditions needed to convert FFA in trap FOG to their corresponding methyl ester as illustrated in Table 5.

The presence of water in trap grease FOG is known to be detrimental to the overall acidic conversion of FFA to methyl esters. Indeed, several publications have noted the need for <1% water in the reaction mixture if the product is to achieve a preferred acid number of 2 mg KOH/g or less. A paper by Montefrio et al. [41] also reported high methyl ester conversion efficiencies for FFA in trap FOG as summarized in Table 6.

TABLE 6: Acid-catalyzed esterification of FOG with varying levels of FFA [46]

% FFA content	Methanol:FFA molar	Final acid value mg KOH/g
20	10.0	8.60
	20.0	1.19
	50.0	0.26
10	10.0	5.24
	20.0	1.18
	50.0	0.26
5	10.0	2.06
	20.0	1.06
	50.0	0.13

H_2SO_4 catalyst, reaction temperature of 30°C, and reaction time of 24 h

They also examined the role of $FeSO_4$ as a catalyst instead of sulfuric acid for esterification of trap FOG FFA as the catalyst is relatively insoluble in a FOG/methanol solution which could facilitate post processing separations but the conversion efficiencies were not as high unless the reaction temperature was elevated to 95°C. Equally important, they examined the impact of stirring on conversion efficiencies which is important as most trap FOG is not soluble in acidic methanol. In the absence of mixing, FFA to methyl ester conversion efficiencies were reported to be in the low 60% range whereas orbital shaking in shake flasks at 350 rpm provided +90% conversion of FFA to FAME. Clearly, for industrial applications further studies will need to be done to optimize high-intensity mixing and one could envisage a future role for surfactant chemistry. A rich source of chemical information for the catalytic conversion of FFA to FAME in trap FOG is the numerous catalytic solid-acid esterification studies that have been accomplished on model compounds and yellow greases [48]. Although there are differences in the nature of the starting materials these prior studies should aid in future catalyst design. A recent notable example was reported by Kim et al. [49] which employed grease interceptor FOG from a Chinese restaurant located in Charlotte, NC. The authors examined the use of H_2SO_4, $NiSO_4/SiO_2$, zeolite, or SiO_2 (0.75–17.5% charge) to convert the FFA in the trap FOG to the corresponding FAME in a methanolic solution. The use of sulfuric acid provided a 99.9% conversion to FAME and $NiSO_4/SiO_2$ a 85% conversion whereas the use of zeolite and SiO_2 was 25% and 18%, respectively.

The second step of transesterifing the mixture of FAME and mono/di/ triacylglycerides to biodiesel has not been as frequently reported, but Karnasuta et al. [39] has provided a detailed central composite design study which is summarized in Table 7.

TABLE 7: Transesterification of methoxylated FOG-yielding biodiesel

Methanol:Oil v/v	KOH concentration % w/v	Reaction Time/h	Free fatty acid methyl ester % yield
0.22–0.30	0.50–1.00	0.45–1.55	84.3–96.1

Optimal conditions for conversion to biodiesel were identified as 0.26 v/v methanol:oil, 1% w/v KOH, for 1 h at 60°C which provided a product with 95.49% fatty acid methyl ester content. A key conclusion from this study was that the resulting biodiesel meet diesel standards for Europe (i.e., EN 14214), China (i.e., GB252-2000), and America (i.e., ASTM D6751).

As an alternative approach to a two-stage treatment, Wang et al. [42] reported a more aggressive methanolysis-utilizing trap grease from Guangzhou Environmental Protection Agency with an acid value of 100 mg/g and 0.8% water. Using methanol and sulfuric acid, they were able to achieve an ester content in the product of 89.67% employing 35.0 methanol:oil ratio and 11.3% acid catalyst stirred for 4.59 h at 95°C. Calcium oxide was then used to neutralize the acid, the calcium sulfate was removed by filtration and the glycerol was decanted. The excess methanol was recovered by a low-temperature distillation and the remaining biodiesel was purified to Korean regulatory biodiesel standards via two distillation towers.

A review of the open literature and web indicates several companies are now actively pursuing the conversion of trap grease FOG to biodiesel although much of the technologies involved are proprietary. The initial processing of trap FOG prior to conversion has not been extensively reported in the research literature but a recent report indicates that prior to esterification of FFA in FOG the trap grease needs to be screened to remove solids, degummed, sulfur depleted, and dried [50]. It also needs to be stated that the conversion of trap grease FOG to biodiesel will be accompanied with a substantial water fraction that needs to be environmentally disposed in an acceptable manner. Hence, cositing this process with a waste water treatment system is one of the most preferred industrial options to pursue. Finally, purification of the FAME/biodiesel is often accomplished via distillation.

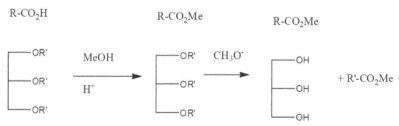

R': H, Fatty Acid-C(O)-
R: Fatty acid saturated/unsaturated

FIGURE 3: Two-stage trap grease fats, oils, and greases conversion to biodiesel and glycerol.

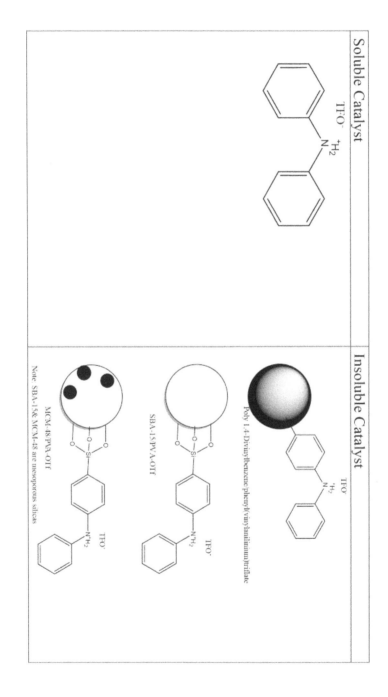

FIGURE 4: Catalyst used for conversion of trap grease fats, oils, and greases (FOG) to FAME. Employed 200 mg trap FOG/experiment; catalyst 0.5–5.0%

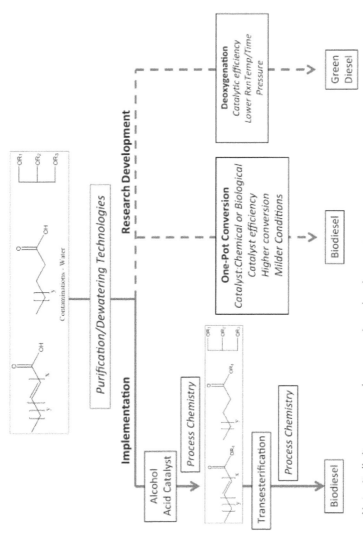

FIGURE 5: Translational research needs in trap fats, oils, and greases to biofuels.

A pilot plant demonstration of converting trap FOG to biodiesel has been reported in detail by Chakrabarti et al. [51] on a 200–400 L. This batch process employed interceptor grease trucked to the East Bay Municipal Utility District (Oakland, CA) after being pretreated by a third party to extract the brown grease. The conversion of the FFA to FAME was accomplished using methanol and H_2SO_4 and if the level of FFA was >1.5–2.0% a second esterification was performed. The esterified FOG was then transesterified with sodium methoxide. The crude biodiesel was water washed, treated with a magnesium silicate absorbent, and filtered. All batches meet the total acid number and glycerin limits for ASTM D6751 method. The most significant challenge was the sulfur limits as the trap FOG feedstock had sulfur levels of 300–400 ppm and ~200 ppm was found in the product. Clearly, this is of concern as recent ultra low sulfur diesel regulations have a 15 ppm sulfur specification for highway diesel fuel (see Federal register, 2001). The researchers were able to address these levels by preforming vacuum distillation and treating with activated charcoal yielding a reported sulfur value of 12 ppm. This methodology was viewed as costly and clearly further research is needed to facilitate S-removal in a more cost effective manner.

Testing the generated biodiesel as B20 or B100 using diesel dump trucks was viewed overall as successful and the drivers were satisfied with the performance although it was mentioned that there was an increased frequency of fuel filter changes with the factors contributing to this not determined. The experimental parameters measured in these trials allowed an evaluation of net energy generation which was reported as 1120 kJ/L-FOG and GHG emission reduction of 0.48 kg CO_2/L-FOG.

7.6 RESEARCH DIRECTED AT IMPROVING FOG TO FAME

Given the environmental and financial interests in converting trap FOG to biodiesel, there are continuing research efforts at improving the overall efficiency and ease of this process. One approach to simplifying the process is to use a solid-acid catalyst which would facilitate acid recovery. Huang et al. [52] studied the use of Amberlyst-15 as a replacement for sulfuric acid. Employing trap grease (KOH value of 100 mg/g and 0.8%

water) with a 27:1 molar ratio of methanol to FFA, and 10% Amberlyst-15 at 95°C for 3 h the conversion of FFA to methyl esters was reported at 98.73%. Equally impressive was the fact that the Amberlyst-15 could be reused for another 10 trials with no loss in activity. These results indicate a substantial improvement over an earlier reported study by Gan et al. [53] that reported conversion efficiencies of 60% with less rigorous conditions.

An interesting alternative solid-acid approach was recently published using magnetic iron-coated nano-sized acid catalyst (iron nanoparticles were coated with poly(glycidyl metharcylate) and sulfonic acid groups on the surface) [54]. This catalyst was shown to readily convert FFA in waste grease (i.e., 0.2 g/experiment from Singapore sewage system, FFA:14% wt, 0.14% moisture) in 96% yield to the corresponding FAME within 2 h at 70°C using a 4% loading of catalyst and a methanol:oil weight ratio 1.4. The authors provided data indicating that this nano-catalyst provided superior performance to Amberlyst-15 which was proposed to be due to the high surface area and ease of accessible to the acid groups of the nano-catalyst. The magnetic property of this catalyst was also shown to help facilitate catalyst recovery using a strong magnetic field.

Along the same approach, Ngo et al. [55] reported a series of homogenous and heterogenous catalysts that were capable of converting trap grease FFA and acyl-glycerides to the corresponding FAME in >95% yield (see Fig. 4). The efficiency of conversion was observed with trap grease samples from Atlanta, GA (0.18% H_2O) and San Francisco, CA (0.27% H_2O) at 125°C for 2 h with 20 molar equivalent of methanol and 0.5–5.0% catalyst. Experimental results did indicate that the homogenous catalyst worked slightly better than the heterogeneous catalyst.

The heterogeneous catalysts were effective in the generation of FAME and also facilitated recovery of the catalyst by simply filtration. As an alternative, several studies have examined using methanol and trap FOG under supercritical conditions. Operating at temperatures of 275–325°C, the supercritical methanol technology is catalyst free and benefits from a homogeneous reaction under these conditions. The drawback of this procedure includes high capital and energy costs and a need for a high molar ratio of methanol to grease [56, 57].

Given the diversity of materials in trap FOG and the presence of water it is only natural to examine alternative biological routes. For example, the application of a whole-cell biocatalyst based on Rhizopus oryzae (ATCC10260) was shown to yield 55% biodiesel from trap FOG (employing 12 mL FOG/experiment) with a 72-h transesterification reaction using no excess methanol in a water-containing system at room temperature [58]. Building on this success, Yan et al. [59] has reported a novel tandem lipase system that is especially tailored to esterifying FFA to FEMA and transesterify acylgylcerides to FAME in one-pot. Using trap FOG grease (2 g/experiment, 21.7% FFA) from Singapore, methanol (1:1 mole ratio to grease) and recombinant *E. coli* wet or dry cells were stirred at 30°C for 6–72 h. Two additional aliquots of methanol (same volume as used at time 0) were added at 6 and 12 h. The recombinant *E. coli* capable of coexpressing *Candida antarctica* and *Thermomyces lanuginosus* lipases was shown to convert the trap FOG grease to biodiesel in 95% yield after 72 h. This approach was shown to be more effective than using individual expressed lipases and furthermore the recombinant *E. coli* cells could be recycled five times and retain 75% productivity. This biological approach in the future could be very attractive given the mild conditions, ease by which the lipases could be generated and well established methodologies for disposal of biological catalysts.

An interesting application of these technological approaches was just demonstrated by Ngo et al. [60] in which he grafted *Candida antarctica* and *Thermomyces lanuginosus* lipases to core-shell structured iron oxide magnetic nanoparticles. Using the same trap grease as Yan et al. used above, the grafted *Candida antarctica* and *Thermomyces lanuginosus* lipases in methanol converted the grease to FAME in 97% and 95% yields, respectively [53]. The bio-grafted catalysts could be readily recovered using an external magnetic field and then recycled.

An alternative approach to the utilization of trap grease is to catalytically deoxygenate FOG molecules yielding a hydrocarbon fuel resource. Toba et al. [61] has reported the use of NiMo, NiW, and CoMo sulfide catalyst for hydrodeoxygenation for silylated trap grease. The former two catalysts showed high and stable hydrogenation activity, whereas CoMo suffered from deactivation. Hydrodeoxygenation was accomplished in +99% at 250–350°C for 3 h pressurized with 7 MPa of hydrogen (0.5 wt catalyst, 10 g FOG/experiment). The product yield of n-paraffin was 94.8% using $NiMo/Al_2O_3$ and 91.4% using $NiMo/B_2O_3$-Al_2O_3 and the remaining components were trace amounts of iso-paraffin and olefins. No alcohols, FFA, or esters were detected in the product mixture. In the long term, this approach may see increasing attention as the need for methanol esterification/transesterification reactions and its post processing is eliminated. Nonetheless, researchers need to identify milder conditions/ improved catalysts so as to accelerate this process into routine process production. Figure 5 provides a general technology map of converting trap grease FOG to a fungible biofuel for today and the future.

As stated in the early part of this review, many governmental organizations have reported the benefits of collecting trap grease FOG before it enters sewer systems and indeed, in many parts of the world, high volume generators of FOG are mandated to collect this material which then needs to be processed in an environmentally acceptable manner. The conversion of this resource into biodiesel today, and possibly green diesel in the future, addresses the green city vision that includes a zero-waste policy that several major metropolitan areas are pursuing [62]. Hence, a confluence of events have propelled the commercialization of trap grease FOG to biofuels and as these technologies are implemented they will certainly benefit from pinch and LCA analysis [63, 64].

In summary, the advent of the green city movement and renewed emphasis on environmental sustainable technologies is providing strong interest in converting societal wastes to value added fuels, chemicals, and materials. Existing mandates for the collection and processing of trap grease FOG makes this a valuable resource for biodiesel production. Although the chemical constituents of trap FOG are variable, general trends are apparent and chemical conversion technologies for the synthesis of biodiesel can be readily accomplished. Current commercial production technologies make this an attractive option and ongoing research will further simplify this process and only accelerate the demand for this valuable resource.

REFERENCES

1. Clark, J. H., R. Luque, and A. S. Matharu. 2012. Green chemistry, biofuels, and biorefinery. Annu. Rev. Chem. Biomol. Eng. 3:183–207.
2. Ragauskas, A. J., C. K. Williams, B. H. Davison, G. Britovsek, J. Cairney, C. A. Eckert, et al. 2006. The path forward for biofuels and biomaterials. Science 311:484–489.

3. Smith, A. M. 2008. Prospects for increasing starch and sucrose yields for bioethanol production. Plant J. 54:546–558.
4. Graham-Rowe, D. 2011. Agriculture: beyond food versus fuel. Nature 474:S6–S8.
5. Foston, M., and A. J. Ragauskas. 2012. Biomass characterization: recent progress in understanding biomass recalcitrance. Ind. Biotechnol. 8:191–208.
6. Ferrell, J., and V. Sarisky-Reed. 2010. National algal biofuels technology roadmap. A technology roadmap resulting from the National Algal Biofuels Workshop Dec. 2008. Available at http://www1.eere.energy.gov/biomass/pdfs/algal_biofuels_road-map.pdf (accessed March 30, 2012).
7. Durbin, T. D., R. Cocker David, A. A. Sawant, K. Johnson, J. W. Miller, B. B. Holden, et al. 2007. Regulated emissions from biodiesel fuels from on/off-road applications. Atmos. Environ. 41:5647–5658.
8. Stelmachowski, M. 2011. Utilization of glycerol, a by-product of the transesterification process of vegetable oils: a review. Adv. Chem. Res. 9:59–92.
9. Feasibility Report Small Scale Biodiesel Production. Available at http://www.wmrc.uiuc.edu/tech/small-scale-biodiesel.pdf (accessed March 30, 2013).
10. Mittelbach, M., B. Pokits, and A. Silberholz. 1992. Production and fuel properties of fatty acid methyl esters from used frying oil. In Liquid Fuels from Renewable Resources: Proc. Alternative Energy Conference. St. Joseph, MI. 74–78.
11. Horton, C. 2007. Removal of free fatty acids from used cooking oil prior to biodiesel production. Brit. U.K. Pat. Appl. GB20060006533 20060331. GB 2436836 A 20071010.
12. Bressler, D. 2011. Methods for producing fuels and solvents substantially free of fatty acids. PCT Int. Appl. WO2011IB00464 20110224. WO 2011104626 A2 20110901
13. Used and Waste Oil and Grease for Biodiesel. Available at http://www.extension.org/pages/28000/used-and-waste-oil-and-grease-for-biodiesel (accessed April 5, 2013).
14. BLACKGOLD Converting our Crudest Wastes into our Cleanest Fuels. Available at http://www.blackgoldbiofuels.com/ (accessed April 5, 2013).
15. FOGFUELS™ Fueling the Future™. Available at http://www.fogfuels.com/ (accessed April 5, 2013).
16. Pacific Biodiesel. Available at http://www.biodiesel.com/ (accessed April 5, 2013).
17. Bacigalupi, R. Turning trap grease, brown grease to biodiesel. Opportunity, process and issues. Available at http://www.gobiomass.com/article.cfm?id=28407&PageNum=2 (accessed April 5, 2013).
18. Bevill, K. 2008. A greasy alternative. Biodiesel Magazine. October 14, 2008.
19. Restaurant Grease: Knowing Your Ohio EPA Regulations. 2010. Available at http://www.epa.ohio.gov/portals/41/sb/publications/restaurant.pdf (accessed April 5, 2013).
20. Weiss, M. 2007. Grease interceptor facts and myths. Plumbing Systems & Designs. Nov., 34- 39.
21. Grease Interceptors A112.14.3 - 2000. Available at http://www.asme.org/products/codes—standards/grease-interceptors (accessed March 30, 2013).
22. Gallimore, E., T. N. Aziz, Z. Movahed, and J. Ducoste. 2011. Assessment of internal and external grease interceptor performance for removal of food-based fats, oil, and grease from food service establishments. Water Environ. Res. 83:882–892.

23. Grease Interceptor Maintenance. Available at http://www.sanjoseca.gov/archives/163/GreaseInterceptorMaintenance_E.pdf; http://www.centralsan.org/documents/Grease_Interceptor_Maint_Fact_Sheet.pdf (accessed April 5, 2013).

24. Understanding Grease Removal And The 25% Rule - City of Dallas. Available at http://www.dallascityhall.com/code_compliance/pdf/GreaseTrap_brochure.pdf (accessed April 5, 2013).

25. Rintoul, S. 2006. Stop FOG now. Water Environ. Lab. Solut. 13:6–8.

26. He, X., M. Iasmin, L. O. Dean, S. E. Lappi, J. J. Ducoste, and F. L. de los Reyes. 2011. Evidence for fat, oil, and grease deposit formation mechanisms in sewer lines. Environ. Sci. Technol. 45:4385–4391.

27. Environmental Protection Agency. 2004. EPA report to congress. Available at http://www.epa.gov/npdes/pubs/csossoRTC2004_chapter04.pdf; Environmental Protection Agency. 2007b. EPA SSO Draft Fact Sheet. Available at http://www.epa.gov/npdes/pubs/sso_fact_sheet_model_permit_cond.pdf (accessed April 5, 2013).

28. Fats, Oils & Grease To Green Fuel, Partanen, W.E. Available at http://www.ncsafewater.org/Pics/Training/AnnualConference/AC08TechnicalPapers/SpecialTopics/AC08ST_Tues1100_Partenan.pdf (accessed April 5, 2013).

29. Fat, Oil and Grease Programme Dublin City Council's FOG Programme. Available at http://www.dublincity.ie/WATERWASTEENVIRONMENT/WASTEWATER/Pages/FatOil andGreaseProgramme.aspx (accessed April 5, 2013).

30. Grease Interceptor Regulation. Available at http://www.metrovancouver.org/SERVICES/PERMITS/GREASETRAPREGULATION/Pages/default.aspx (accessed April 12, 2013).

31. Grease Under Control at South East Water, Scoble, C., Day, N. 2002. Proceedings of the 65th Annual Water Industry Engineers and Operators' Conference. Available at http://www.wioa.org.au/conference_papers/02/paper7.htm (accessed April 5, 2012).

32. Suto, P., D. M. D. Gray, E. Larsen, and J. Hake. 2006. Innovative anaerobic digestion investigation of fats, oils, and grease. Proc. Water Environ. Fed. Residuals Biosolids Manage. 22:858–879.

33. Long, J. H., T. N. Aziz, F. L. III de los Reyes, and J. J. Ducoste. 2012. Anaerobic co-digestion of fat, oil, and grease (FOG): a review of gas production and process limitations. Process Saf. Environ. Prot. 90:231–245.

34. Sheng, L. 2012. 'Gutter oil' sewer problem drained. Global Times. Available at http://www.globaltimes.cn/content/714805.shtml (accessed April 5, 2012).

35. Wiltsee, C. 1998. Urban Waste Grease Resource Assessment. Available at http://www.epa.gov/region9/waste/biodiesel/docs/NRELwaste-grease-assessment.pdf (accessed April 5, 2012).

36. Wiltsee, G. 1998. Urban waste grease resource assessment. NREL/SR-570-26141, Nov.

37. Canakci, M. 2007. The potential of restaurant waste lipids as biodiesel feedstocks. Bioresour. Technol. 98:183–190.

38. Nisola, G. M., E. S. Cho, H. K. Shon, D. Tian, D. J. Chun, E. M. Gwon, et al. 2009. Cell immobilized FOG-trap system for fat, oil, and grease removal from restaurant wastewater. J. Environ. Eng. 135:876–884.

39. Karnasuta, S., V. Punsuvon, C. Chiemchaisri, and K. Chunkao. 2007. Optimization of biodiesel production from trap grease via two-step catalyzed process. Asian J. Energy Environ. 8:145–168.

40. Neczaj, E., J. Bien, A. Grosser, M. Worwag, and M. Kacprzak. 2012. Anaerobic treatment of sewage sludge and grease trap sludge in continuous co-digestion. Global Nest J. 14:141–148.

41. Montefrio, M. J., T. Xinwen, and J. P. Obbard. 2010. Recovery and pre-treatment of fats, oil and grease from grease interceptors for biodiesel production. Appl. Energy 87:3155–3161.

42. Wang, Z. M., J. S. Lee, J. Y. Park, C. Z. Wu, and Z. H. Yuan. 2008. Optimization of biodiesel production from trap grease via acid catalysis. Korean J. Chem. Eng. 25:670–674.

43. Coker, C. 2006. Composting grease trap wastes. Biocycle 47:27–30.

44. Robbins, D. M., O. George, and R. Burton. 2011. Developing programs to manage fats, oil, and grease (FOG) for local governments in India. V. World Aqua Congress Proceedings, New Delhi, India, 1–13.

45. Ward, P. M. L. 2012. Brown and black grease suitability for incorporation into feeds and suitability for biofuels. J. Food Prot. 75:731–737.

46. Vasudevan, P. T., and B. Fu. 2010. Environmentally sustainable biofuels: advances in biodiesel research. Waste Biomass Valorization 1:47–63.

47. Perego, C., and D. Bianchi. 2010. Biomass upgrading through acid-base catalysis. Chem. Eng. J. 161:314–322.

48. Diaz, L., M. E. Borges, 2012. Low-Quality Vegetable Oils as Feedstock for Biodiesel Production Using K-Pumice as Solid Catalyst. Tolerance of Water and Free Fatty Acids Contents. J. Agric. Food. Chem. 60:7928–7933.

49. Kim, H. J., H. Hilger, and S. J. Bae. 2013. NiSO4/SiO2 catalyst for biodiesel production from free fatty acids in brown grease. Energy Eng. 139:35–40.

50. Bacigalupi, R. 2010. Turning trap grease, brown grease to biodiesel. Opportunity, process and issues. Biomass Products & Technology, Aug.

51. Chakrabarti, A. J., J. M. Hake, I. Zarchi, and D. M. D. Gray. 2008. 4Waste grease biodiesel production at a wastewater treatment plant. Water Environ. Fed. ???:2770–2789.

52. Huang, D., Z. Wang, and Z. Yuan. 2012. Pretreatment of trap grease with Amberlyst-15 for biodiesel production. Adv. Mater. Res. 347–353:2528–2531.

53. Gan, S., H. K. Ng, P. H. Chan, and F. L. Leong. 2012. Heterogeneous free fatty acids esterification in waste cooking oil using ion-exchange resins. Fuel Process. Technol. 102:67–72.

54. Zillillah, T. G., and Z. Li. 2012. Highly active, stable, and recyclable magnetic nanosize solid acid catalysts: efficient esterification of free fatty acid in grease to produce biodiesel. Green Chem. 14:3077–3086.

55. Ngo, H. L., Z. Xie, S. Kasprzyk, M. Haas, and W. Lin. 2011. Catalytic synthesis of fatty acid methyl esters from extremely low quality greases. J. Am. Oil Chem. Soc. 88:1417–1424.

56. Alsoudy, A., M. Hedeggard, T. Janajreh, and I. Janajreh. 2012. Influence on process parameters in transesterification of vegetable and waste oil – a review. Int. J. Res. Rev. Appl. Sci. 10:64–77.

57. Preprocessinc Grease Trap Waste to Feedstock to Fuel. Available at www.prepro-cessinc.com/…/b0351bd5e310031f 4639473c97ed0abe.pdf; http://www.extension.org/pages/28000/used-and-waste-oil-and-grease-for-biodiesel (accessed April 5, 2012).

58. Jin, G., T. J. Bierma, C. G. Hamaker, R. Mucha, V. Schola, J. Stewart, et al. 2009. Use of a whole-cell biocatalyst to produce biodiesel in a water-containing system. J. Environ. Sci. Health A Tox. Hazard. Subst. Environ. Eng. 44:21–28.

59. Yan, J., A. Li, Y. Xu, T. P. N. Ngo, S. Phua, and Z. Li. 2012. Efficient production of biodiesel from waste grease: one-pot esterification and transesterification with tandem lipases. Bioresour. Technol. 123:332–337.

60. Ngo, T. P. N., A. Li, K. W. Tiew, and Z. Li. In press. Efficient transformation of grease to biodiesel using highly active and easily recyclable magnetic nanobiocata-lyst aggregates. Bioresour. Technol.

61. Toba, M., Y. Abe, H. Kuramochi, M. Osako, T. Mochizuki, and Y. Yoshimura. 2011. Hydrodeoxygenation of waste vegetable oil over sulfide catalysts. Catal. Today 164:533–537.

62. Ferry, D. 2011. Urban Quest 'Zero' Waste Wall Street J. September 11. Available at http://online.wsj.com/article/SB10001424053111904583204576542233226922972.html (accessed April 5, 2013).

63. Ebrahim, M., and A. Kawari. 2000. Pinch technology: an efficient tool for chemical-plant energy and capital-cost saving. Appl. Energy 65:45–49.

64. Majer, S., F. Mueller-Langer, V. Zeller, and M. Kaltschmitt. 2009. Implications of biodiesel production and utilisation on global climate – a literature review. Eur. J. Lipid Sci. Technol. 111:747–762.

CHAPTER 8

Efficient Extraction of Xylan from Delignified Corn Stover Using Dimethyl Sulfoxide

JOHN ROWLEY, STEPHEN R. DECKER, WILLIAM MICHENER, AND STUART BLACK

8.1 INTRODUCTION

Biofuels are becoming more widespread throughout the United States as more advanced conversion methods become available. The most advanced process currently is the conversion of lignocellulosic biomass into ethanol (Kim et al. 2009). Despite having much larger production potential than starch-based ethanol, lignocellulosic ethanol is still in the early stages. The conversion of biomass sugars into biofuels is an important aspect of the Department of Energy's mission to promote the integration of renewable fuels and is a key component in the worldwide move towards renewable energy. Before additional progress can be made, it is desirable to understand in detail the mechanisms that occur during the biomass to biofuel conversion process.

Efficient Extraction of Xylan from Delignified Corn Stover Using Dimethyl Sulfoxide. © *Rowley J, Decker SR, Michener W, and Black S.* 3 Biotech *3,5 (2013), doi: 10.1007/s13205-013-0159-8. Licensed under Creative Commons Attribution License, http://creativecommons.org/licenses/by/2.0/.*

Biomass is made up of three components: cellulose, hemicellulose and lignin. Xylan, a prevalent plant cell wall polymer made up of mostly xylose, is of particular interest as the dominant plant cell wall hemicellulose (Ebringerová et al. 2005). One of the challenges associated with the efficient production of biofuels involves the selective removal and/ or hydrolysis the polymeric xylose backbone of xylan. During neutral or acidic thermochemical pretreatment of biomass, xylan is removed from the biomass and broken down into xylose, arabinose, and a few other minor components such as acetic acid (Naran et al. 2009).

To better understand the mechanism of thermochemical and enzymatic removal of xylan, it is useful to develop antibodies capable of tagging xylan in biomass. Antibodies can be tagged with fluorescent dyes, allowing the location of the xylan in biomass to be tracked either optically or spectrophotometrically prior to and following pretreatment. By identifying the location of the xylan, the pretreatment process and the subsequent fermentation process can be tailored to improve ethanol production. Antibody tagging can be very beneficial in understanding the mechanism of xylan removal, however, to create specific antibody tags, a native-like xylan is desirable. Many extraction methods result in degradation or deacetylation of the xylan resulting in a non-native, water-insoluble product, which could potentially produce antibodies with non-useful specificity, as specific side groups are missing. Dimethyl sulfoxide (DMSO) extractions have been found to result in a water-soluble form of xylan, which retains the acetyl groups present in the native state (Hägglund et al. 1956). This native-like xylan is more likely to result in production of antibodies specific to the native structures found in xylans in situ in the cell wall.

In this study, a DMSO extraction of xylan in corn stover was studied at varying temperatures of extraction to determine an ideal temperature for efficient extraction.

When extracting xylan from biomass with DMSO, a pretreatment of the sample is necessary to open the cell structure and allow the polymeric xylans freedom to be extracted. Owing to the coupling between xylan and lignin, xylan is intractable until much of the lignin has been removed or these connections severed. Decoupling of xylan from lignin is important in accessing xylan in biomass, but complete removal of lignin will result in loss of xylan from the sample (Ebringerová et al. 2005). Multiple delignification procedures exist for the removal of lignin from corn stover, however, acid-chlorite bleaching was found to be the most efficient method of delignification without excessive de-acetylation of the xylan (Ebringerová et al. 2005).

Following delignification, xylan is extracted from the sample. Often xylan is extracted with KOH or NaOH (Ebringerova and Heinze 2000). However, this method results in de-esterification of the acetyl groups present on the xylan (via saponification of the ester links), leading to a water insoluble product which has limited utility for antibody production and as a substrate for hemicellulase assays. Therefore, in this study, xylan was removed by DMSO extraction to retain the acetyl groups, resulting in a water-soluble product. The extraction was first performed at room temperature, following the method proposed by Hägglund et al. (1956) in 1956. This method is carried out by stirring the biomass in DMSO for approximately 24 h at room temperature. A series of extractions was then performed at higher temperatures (70 °C and at 40 °C) with variable times of extraction. The yields resulting from the extractions were compared and, including the time required to perform each extraction, the most efficient method of extraction was determined.

Further analysis was performed on each sample to determine the content of the yield acquired through extraction and to ensure that no significant structural changes took place under heated conditions. Infrared spectroscopy and QToF MS analysis was used to determine the general structural features and to ensure that no de-esterification or de-polymerization took place during the heated extractions.

8.2 METHODS

8.2.1 DELIGNIFICATION OF BIOMASS

Approximately 300 g of milled corn stover was extracted in a polypropylene thimble using a Soxhlet extractor following NREL's Determination of Biomass Extractives Laboratory Analytical Procedure (Sluiter et al. 2008). The NREL procedure is a two-step procedure carried out in a Soxhlet extractor. All extractions are carried out at the reflux temperature of the solvent used and at ambient pressure. Each extraction is performed until little to no color is present in the extraction chamber. Depending on the nature of the material, this takes between 18 and 48 h for each step. The first extraction was performed with de-ionized (DI) water to remove accessible water-soluble compounds. A second extraction was performed using ethanol to remove lipids and other extractables. The solid sample was air-dried following ethanol extraction prior to delignification.

Delignification was carried out in double bagged one gallon plastic zipper closure bags by adding water to the approximately 100 g of air-dried, extracted biomass at a biomass/water consistency of 10 %. Approximately, 40 g of sodium chlorite ($NaClO_2$) was added to the mixture and the bag was mixed well followed by a 5 mL addition of concentrated hydrochloric or glacial acetic acid. A smaller volume of hydrochloric acid is needed to sustain the reaction. The bag was closed and heated in a 60 °C water bath in a fume hood for approximately 3 h. Regular venting of the bag was required to relieve pressure in the bag and prevent reaching too high a concentration of ClO_2. If the concentration of chlorine dioxide in the atmosphere of the bag or any bleaching vessel is too high, a "puff" can result from the decomposition of the chlorine dioxide. A "puff" is a term coined within the pulping industry to differentiate a low speed detonation wave of <1 m/s from an explosion wave (>300 m/s). Plastic zipper bags will open in the event of a puff releasing the gas without creating a debris hazard (Fredette 1996).

Once every hour, an additional 40 g of $NaClO_2$ was added to the bag until the total amount was approximately 0.70 g $NaClO_2$/g biomass. The remaining liquid was filtered from the solids and the solid biomass was thoroughly washed with DI water and lyophilized prior to DMSO extraction.

8.2.2 DMSO EXTRACTION

A 1 L electrically heated reaction flask fitted with an overhead mechanical stirrer was used for all extractions. Approximately 50 g of delignified corn stover was added to a flask and extracted with DMSO using a ratio of approximately 14 mL/g biomass at room temperature with stirring at 20 rpm for a specified time. The solid was filtered and extracted a second time with

DMSO for the same time period. The solid was filtered and washed thoroughly with ethanol to remove residual DMSO and extracted xylan. The ethanol filtrate was reserved for the precipitation step. The DMSO extracts were combined and absolute ethanol was added to the DMSO extract (3.8 L ethanol/L of final extract). Concentrated hydrochloric acid (HCl) was added in a ratio of approximately 0.66 mL HCl/L of ethanol/DMSO solution to precipitate the xylan from the DMSO/ethanol mixture. The solution was cooled at 4 °C overnight to complete precipitation. The cold solution was filtered though paper filter (Whatman Grade 1). The filter paper and isolated xylan were macerated, washed with ethanol and stirred overnight in a small amount of ethanol. Ethanol was filtered from the solid xylan and macerated paper filter. The resulting filter cake was stirred overnight with fresh ethanol to remove as much DMSO as possible and filtered. The filter cake was further washed with diethyl ether with overnight stirring to remove any remaining ethanol and DMSO. The xylan was dissolved away from the macerated paper fibers in warm water (30 °C), filtered with small amounts of water added for washing and lyophilized.

The DMSO extraction was carried out at 20, 40 and 70 °C according to the conditions shown in Table 1. Extractions at 40 and 70 °C were performed in duplicate. Extraction at 20 °C was a single extraction.

8.2.3 SAMPLE ANALYSIS

The final products were analyzed qualitatively by their water solubility and for yield from the bleached material by mass. The methods were compared according to yield and time efficiency. Samples were analyzed on a Thermo Scientific Nicolet 6700 FTIR Spectrometer fitted with a Smart iTR diamond cell and a DTGS detector. Samples were scanned for 150 scans and compared to previously isolated and analyzed samples (Ebringerová et al. 2005).

Two samples, one extracted at room temperature, and the other at 70 °C, were prepared in a 50/50 solution of H_2O/acetonitrile in 0.2 % formic acid. Each sample was directly infused into a Micromass Q-ToF micro (Micromass, Manchester, UK) quadrupole time of flight mass spectrometer with a 250 μL Hamilton gastight syringe (Hamilton, Reno, NV, USA) at a flow rate of 5 μL/min. Spectra were obtained in positive MS mode from a mass range of 600–1,500 m/z and processed by Masslynx data system software (Micromass, Manchester, OK). In positive-ion MS mode, cone voltage was set at 30 volts and capillary at 3,000 volts. Both cone and desolvation gas flows were optimized at 10 and 550 L/h, respectively. Source and desolvation temperatures were set at 100 and 250 °C. Sample mass spectra were collected for 2 min to ensure adequate signal levels. Mass calibration was performed using a solution of 2 pmol/μL of sodium rubidium iodide solution. The calibration mix was collected for 2 min and summed.

TABLE 1: The conditions for four subsequent extractions at temperatures above room temperature

Temperature (°C)	Time (h)	Number of extractions
70	2	2
70	2	1
70	1	2
40	4	2
20	24	2

The time noted above represents the duration of each extraction

8.3 RESULTS

Figure 1 shows the percent yield for each of the extracted samples. Analysis of the yield for the different methods indicates that heating during the extraction process results in no significant loss of recovery. It is also evident that much less time is needed for extraction when the DMSO is heated during extraction compared to a room temperature extraction. At room temperature, two sequential extractions, each lasting a full 24 h, are necessary and resulted in a xylan/delignified biomass yield of 8.7 %. Upon heating to 70 °C and decreasing the extraction time to only 2 h, the yield was 8.6 ± 0.2 %. Even when the extraction time was decreased to 1 h for a sample heated to 70 °C, the loss in yield was not found to be particularly large (7.6 ± 0.6 %). However, a significant loss in yield was found when the sample was extracted only once with DMSO. The percent yield did not drop significantly when the sample was heated to 40 °C, but a longer extraction time was necessary (Fig. 1).

By infrared spectroscopy, there is little structural difference between the heated and room temperature extractions. Figure 2 compares corn stover xylan extracted using DMSO and a commercial oat spelt xylan (Fluka) extracted under alkaline conditions. The commercial xylan has no signal for the acetate ester present in the DMSO xylans at ~1,700 and ~1,300/cm which are the well-known carbonyl and ester linkage absorbance bands for the acetyl groups. The commercial xylan shows a slight absorbance at 1,500/cm which is indicative of residual lignin. The DMSO extracted lignin does not have an absorbance in this region. This would indicate that no either no residual lignin was present following acid chlorite delignification or that no water soluble lignin was present in the isolated xylan following lyophilization.

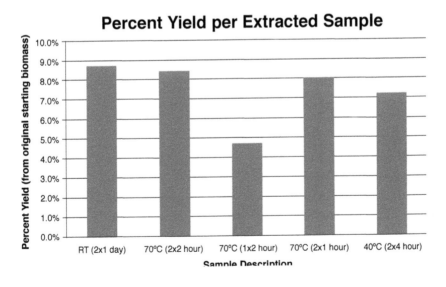

FIGURE 1: % yield of xylan extracted from bleached material. Descriptors of each sample are temperature (# replicates × time of extraction for each replicate); RT room temperature

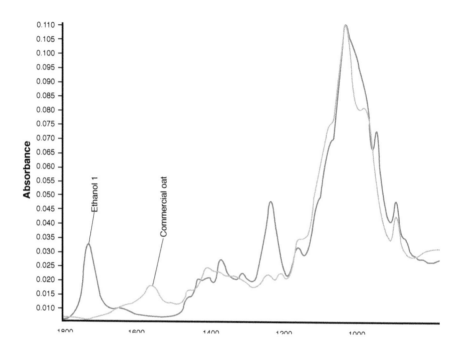

FIGURE 2: IR-spectra of xylan samples. The red spectrum is of the DMSO extracted xylan. The green spectrum is of a commercial oat xylan sample extracted with alkaline conditions. The peaks at 1,735 and 1,235/cm are indicative of carbonyl and ester linkage, respectively, of the acetyl groups on the xylan polymer

When comparing (Fig. 3) the two DMSO extracted corn stover xylans, it is clear that heating during the DMSO extracting process does not influence the structure in a significant way. The expected peaks are present for the DMSO extracted sample indicating the presence of an ester group.

Both the room temperature- and 70 °C-extracted samples were water soluble, providing further evidence of the presence of acetyl groups on the isolated xylans, as acetylation is known to provide for water solubilization of xylans (Grondahl and Gatenholm 2005; Gabrielii et al. 2000). Figure 4 shows the mass spectra comparison between xylan extracted at room temperature and xylan extracted at 70 °C. The MS spectra collected from each

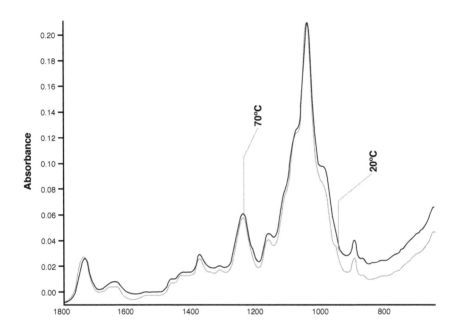

FIGURE 3: IR-spectra of two extracted xylan samples. The blue line is the spectrum of the DMSO extracted sample extracted at 70 °C. The red line is the spectrum for a corn stover extracted xylan at room temperature

sample shows a degree of polymerization range of 4–9 residues, indicative of the limitations of the ESI technique, rather than the actual DP of the sample materials. Low MW xylo-oligomer standards (DP 2–4) showed no fragmentation at the voltages used in this study (data not shown), indicating that the ions detected in the samples are generated during the sample preparation and remain with the soluble fraction during purification, not as artifacts of MS fragmentation. There are slight differences in the two spectra, specifically with intensities seen at varying masses with the relative abundance. The spectra show that a high number of the masses associated with each sample are present in the other. It is clear from the fragments in the spectra that the two xylan samples are structurally very similar, supporting the analysis from the IR instrument, and showing that no significant structural changes occurred when xylan was extracted at a higher temperature. Elucidation of the structure of the isolated xylans may be found in our previous work (Naran et al. 2009) and was not attempted in this study, which is primarily aimed at developing a faster, easier method for obtaining native xylans.

8.4 DISCUSSION

Drawing from these results, it can be concluded that heating during an extraction can increase the efficiency of a xylan extraction. A heated extraction requires much less time than an extraction done at room temperature. It also must be mentioned that while further study is needed, these preliminary results predict that the number of extractions (proportional to the total volume of DMSO used) does impact that percent yield. The conclusion that percent yield is increased upon heating is not supported by this study, however, further analysis can be done to confirm this prediction. These results strongly indicate that the yield obtained under heated conditions is comparable to that of an unheated extraction and requires significantly less time to extract (~9 % of the total time required to extract an unheated sample). Further, it must also be said that heating during the extraction process within the temperature range studied here does not change the xylan structure. The product is not de-esterified during the process and remains water soluble. Furthermore, no significant or obvious structural changes were observed when comparing heated samples to non-heated samples (Figs. 3, 4).

The efficiency of xylan extraction with DMSO can be greatly improved if the samples are heated during extraction. The percent yield resulting from an extraction performed at 70 °C for 2 h with multiple extractions is comparable to the yield at room temperature with two 24 h extractions. Provided that no de-esterification of the xylan results from heating the sample as shown in Fig. 3, heating increases the efficiency of the extraction. From this study, it was determined that the most efficient method of extraction is the following: two 70 °C extractions each lasting 2 h. This method provided a yield of 8.6 ± 0.2 % which is considered to be sufficient for the purposes of this study.

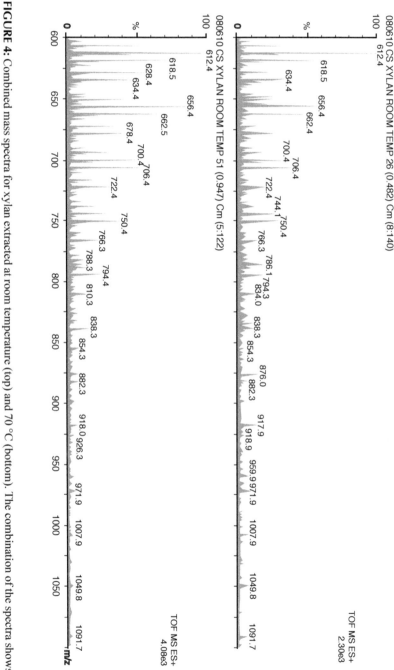

FIGURE 4: Combined mass spectra for xylan extracted at room temperature (top) and 70 °C (bottom). The combination of the spectra shows that the two samples are structurally the same

REFERENCES

1. Ebringerova A, Heinze T (2000) Xylan and xylan derivatives—biopolymers with valuable properties, 1—naturally occurring xylans structures, procedures and properties. Macromol Rapid Comm 21(9):542–556. doi:10.1002/1521-3927(20000601)21:9<542::Aid-Marc542>3.3.Co;2-Z
2. Ebringerová A, Hromadkova Z, Heinze T (2005) Hemicellulose. Adv Polym Sci 186:1–67
3. Fredette MC (1996) Bleaching chemicals: chlorine dioxide. In: Reeve DW, Dence CW (eds) Pulp bleaching: principles and practice. Tappi Press, Atlanta, pp 67–68
4. Gabrielii I, Gatenholm P, Glasser WG, Jain RK, Kenne L (2000) Separation, characterization and hydrogel-formation of hemicellulose from aspen wood. Carbohydr Polym 43(4):367–374. doi:10.1016/S0144-8617(00)00181-8
5. Grondahl M, Gatenholm P (2005) Role of acetyl substitution in hardwood xylan. In: Dumitriu S (ed) Polysaccharides: structural diversity and functional versatility. Marcel Dekker, New York, pp 509–514
6. Hägglund E, Lindberg B, McPherson J (1956) Dimethylsulphoxide, a solvent for hemicelluloses. Acta Chem Scand 10:1160–1164
7. Kim TH, Nghiem NP, Hicks KB (2009) Pretreatment and fractionation of corn stover by soaking in ethanol and aqueous ammonia. Appl Biochem Biotechnol 153(1–3):171–179. doi:10.1007/s12010-009-8524-0
8. Naran R, Black S, Decker SR, Azadi P (2009) Extraction and characterization of native heteroxylans from delignified corn stover and aspen. Cellulose 16(4):661–675. doi:10.1007/S10570-009-9324-Y
9. Sluiter A, Ruiz R, Scarlata C, Sluiter J, Templeton D (2008) Determination of extractives in biomass. National Renewable Energy Laboratory, Golden

CHAPTER 9

The Possibility of Future Biofuels Production Using Waste Carbon Dioxide and Solar Energy

KRZYSZTOF BIERNAT, ARTUR MALINOWSKI, AND MALWINA GNAT

9.1 INTRODUCTION

The Earth's energy requirements are estimated at 14 TW/ y. Considering the economic development, and therefore high consumption and constantly increasing number of people in the world, it is estimated that energy demand in 2050 will be amount 28–30 TW/ y. Fuels from crude oil supply about 96% of the worldwide energy demand for transport. On the other hand, known petroleum reserves are limited and will eventually run out. According to preliminary calculations, fossil fuels will be exhausted within 150–200 years. Fuel consumption causes the emission of carbon dioxide into the atmosphere, resulting in the collapse of the balance between carbon dioxide released to environment, and gas that can be absorbed by plants. It is estimated, that in case of continued use of traditional energy sources by 2030, carbon dioxide levels will rise to 40 billion Mg per year.

The correlation of carbon dioxide emissions from the world's population is shown in Figure 1.

Global emissions of CO_2 and other GHGs, despite commitments of the reduction made by developed countries, will continue to grow, because of increasing production in developing countries, for which clean technologies and investing in renewable energy sources are too expensive. Currently, such a trend can be observed, because the developed countries carbon dioxide emissions was reduced by 6.5% (according to IEA [2] data for 2009). On the other hand the developing countries increased the emissions up to 3.3% (mainly in Asia and the Middle East). In this way, the twenty-first century economy will be depend on fossil fuel resources. As a consequence of this state, GHG concentrations will increase, resulting in the continued progress of global warming. Club of Rome, in 1972, has presented a report: "The Limits to Growth" which predicted that before 2072, present industrial civilization will collapse, as a result of the lack of available energy resources, or because of polluted environment.

It is important to get alternative energy sources, that will be able increasingly replace the fossil fuels with reducing effect of carbon dioxide emissions at the same time. That may be renewable energy sources (RES), such as wind, water, geothermal or solar energy. The potential of the solar

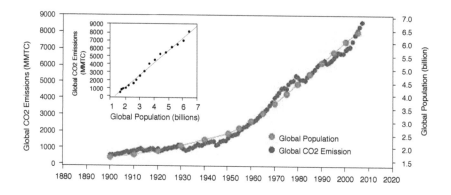

FIGURE 1: The correlation of CO_2 emissions with the world's population [1]

energy is estimated up to 100,000 TW/y. This huge amount of energy has high potential of application in thermochemical biomass conversion or artificial photosynthesis for processing carbon dioxide and water into the organic compounds.

9.2 SOLAR RADIATION AS AN ENERGY SOURCE

Solar energy is the most important source of the energy used on the Earth. According with hypothesis of H. Bethe and C. Weizsacker made in 1938— the energy of the Sun, has its source in the fusion reactions that occur in the interior, according to the reaction (1): [3]

$$4 \ 1H \rightarrow 4He + 2e+ + 2ve + E \tag{1}$$

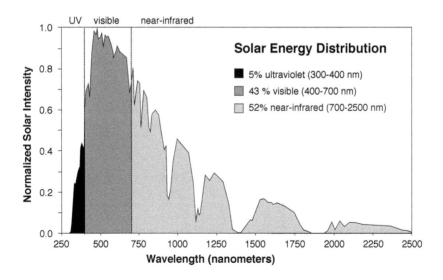

FIGURE 2: The spectrum of solar radiation [5]

Solar radiation is in the form of a wide band of the electromagnetic spectrum, which is shown in Figure 2. It covers wavelengths from about 250 nm to 1000 nm over [4].

However, the full spectral range of solar radiation (including the ultraviolet (UV)) reaches only to the edge of Earth's atmosphere. The total energy that reaches the Earth's atmosphere is marked as Fs—a stream of sunlight that the average power is 1368 W/m². Some of the radiation wavelength, passing through the layers of the Earth's atmosphere, are absorbed by the molecules. Even at the height of the Kármán line (about 100 km above sea level) there is the first interaction of solar radiation with occurring nitrogen and oxygen. Ultraviolet (UV), at a wavelength below 280nm, is high-energy radiation. The energy is enough to cause dissociation of molecules into atoms as is shown in following equations 2 and 3:

$$N_2 + hv \ (\lambda \approx 126nm) \rightarrow 2N \quad \Delta H° = 945 \ kJ/mol \tag{2}$$

$$O_2 + hv \ (\lambda \approx 240nm) \rightarrow 2O \quad \Delta H° = 498 \ kJ/mol \tag{3}$$

In the lower layer of the atmosphere—called the ionosphere—solar energy is absorbed by the reactions occurring in the ionization of chemical individuals (reactions 4 and 5).

$$N_2 + hv \ (\lambda \approx 80nm) \rightarrow 2N++ e \quad \Delta H° = 1500 \ kJ/mol \tag{4}$$

$$O + hv \ (\lambda \approx 91nm) \rightarrow O++ e \quad \Delta H° = 1310 \ kJ/mol \tag{5}$$

Moving towards next layer—ozonesphere—sunlight meets molecules of ozone (O_3). In this way the next part of the radiation is absorbed by the ozone, which leads to O_3 dissociation, resulting in the formation of excited molecules and atoms of oxygen (reaction 6):

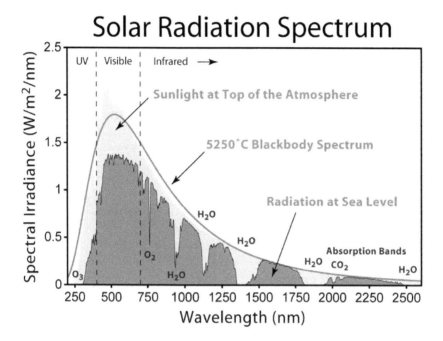

FIGURE 3: Sunlight spectrum before and after its passage through the atmosphere of the Earth [6]

$$O_3 + hv \ (\lambda < 325nm) \rightarrow O_2{}^* + O^* \qquad (6)$$

In addition, a small amount of solar radiation is absorbed during its passage through the troposphere.

As a result, after the absorption and dispersion in the atmosphere, the spectral range of solar radiation flux reaching the Earth's surface is slightly changed, mostly free of long-range radiation from ultraviolet as seen in Figure 3.

By assuming the all solar energy which reaches the Earth's atmosphere amount 100 units, as many as 19 of them will be absorbed by molecules and suspensions occurring in the Earth's upper atmosphere. Other relative flows of solar radiation towards the Earth shows Figure 4.

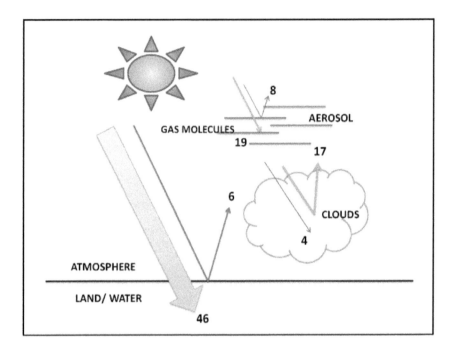

FIGURE 4: The flow of solar radiation reaching the Earth's surface [4 modified]

As shown in Figure 4, part of the solar radiation in general does not take part in the energy balance of the Earth—it is reflected to the space. 31 units is reflected including: 6 units directly reflected by the surface of the Earth, 8 units is reflected by aerosols, dust and other materials such as volcanic ash, and the remaining 17 units are reflected from the clouds. The value of solar radiation is expressed as a percentage by the number of Albedo. According to the calculations the Earth's Albedo amounts 31%. From the remaining part of the solar energy (50 units)—4 units are still absorbed by clouds of drops suspended in the water, in consequence remains 46 units. In this way only 46% of solar radiation reaches the Earth's surface and can be absorbed by the land and water. Radiation— a relatively high-and shortwave (UV and VIS)—is partially stored and used in different processes on the Earth, and part of it is reemitted into space in the form of long-wave radiation (infrared, IR) with less energy.

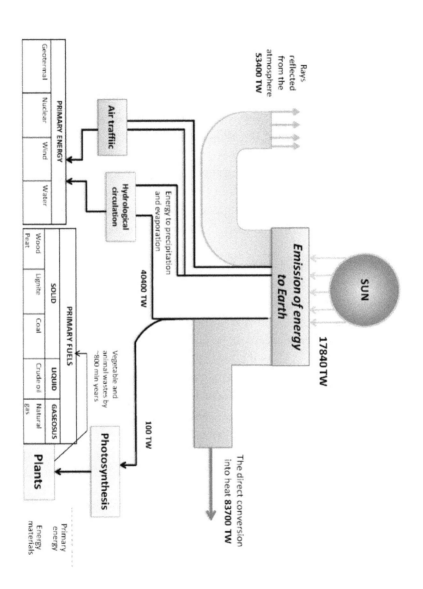

FIGURE 5: Diagram of solar energy transformations [7 changed]

Therefore, the total amount of energy absorbing solar radiation reaching the earth shield can be expressed by equation 7:

$$E_s = F_s(1 - A)\, \pi r2 \tag{7}$$

where: E_s means the total amount of solar energy which is absorbed by the Earth (W), F_s—average flux of solar radiation (solar constant) (1368 W/m^2), A—(albedo) of the radiation reflected back into space (0.31), r—the radius of the Earth (6.37 106).

Therefore, the absorbed solar energy is approximately 1.2 10^{17}W (about 124,600 TW) [7]. This amount of absorbed solar radiation is the driving force behind all the changes that are taking place on the globe. The solar energy transformation scheme, that reaches the Earth's surface is shown in Figure 5.

Most of the solar energy stream is directly converted into the heat, which is about 67% (nearly 83,700 TW) of the radiation reaching the Earth's surface. About 32% of this energy take part in the hydrological Earth cycle (about 40,400 TW). The energy of water (oceans, seas, inland waters), tidal wave energy can be used to produce electricity in hydroelectric power plants such as the, so-called, flow power plants, which is based on a natural flow. Part of the solar energy is converted into kinetic energy of the wind. Around 400 TW is used for air movement. Wind as an energy source was used in ancient times, and today the kinetic energy of winds is used to produce electricity by using wind turbines.

Other 100 TW of solar radiation are driving forces behind the production of biomass. That organic matter is produced from the processing of solar energy through photosynthesis.

9.3 THE PROCESSES OF PHOTOSYNTHESIS AND THEIR IMPACT ON THE BIOMASS GROWTH

Green plants, some bacteria and protists have developed specific mechanism for synthesis of reduced carbon compounds, through which energy from the sun has been successfully transformed into a useful form of it. This process, which is one of the most important biochemical processes on the Earth, is called photosynthesis. Its name comes from two ancient Greek words meaning "light" and "connect". The total formula for process of photosynthesis is a reaction combining water molecules and carbon dioxide, in the presence of energy from the sun, to give the product as a basic sugar molecule, and oxygen as a byproduct (reaction 8).

$$nH_2O + nCO_2 \rightarrow C_nH_{2n}O_n + nO_2 \tag{8}$$

If assumed the final product of photosynthesis is glucose, simple sugar molecule belongs to a group of a hexoses. Then the total reaction formula can be written as below (Reaction 9):

$$6H_2O + 6CO_2 \rightarrow C_6H_{12}O_6 + 6O_2 \tag{9}$$

Photosynthesis can be distinguished by two sets of reactions: the light-dependent reaction and light-independent reaction (also called dark reactions of photosynthesis). The photosynthesis is initiate by solar radiation, falling on the surface of green plant leaves, is absorbed by assimilation pigments, acting as a catalyst [4]. There are two types of these pigments: chlorophyll a and chlorophyll b. Chlorophyll molecules have strong absorption properties, absorbs solar energy from the electromagnetic spectrum in the range of 400 nm to 700 nm. Radiation of this wavelength range is called Photosynthetically Active Radiation (PAR). The absorption spectrum of chlorophylls differing slightly from each other. Chlorophyll a with total formula $C_{55}H_{72}O_5N_4Mg$ have blue-green color and absorbs light violet wavelength of 417 nm, and the red one wavelength of 657 nm. Chlorophyll b ($C_{55}H_{72}O_5N_4Mg$) absorbs blue light in the field of 460 nm and red one 650 nm wavelength [8]. These small shifts of the both colors absorption maximums is the result of a small difference in the construction. The methyl group present in the chlorophyll a has been replaced by an aldehyde group in the molecule of chlorophyll b (Figure 6). The Range of the spectrum, useful for photosynthesis, is enhanced by auxiliary pigments—carotenoids, which have absorption properties of solar energy, not available for green chlorophyll. The color of the carotenoids is yellow-orange, which is the result of absorption by the blue-violet wavelength of 400 nm-495 nm (Figure 7) [9].

Chlorophylls and carotenoids can be found in chloroplasts tylakoids. Dyes are arranged briefly to form units called photosystem I and photosystem II. Visible light beam falling on the chlorophyll molecule, starts the electron excitation state of the magnesium ion at the center of the porphyrin ring. Excited electron is transferred into a conjugated bonds, and then transported to the neighboring molecule dyes. Thus, to pass the initial electron acceptor in photosystem II (which is ferrodoksin molecule) to photosystem I (where particles of quinone are primary acceptor) that ends on the molecule of coenzyme NADP + and reduce it to NADPH. Chlorophyll molecule, after return the electron, has strong oxidizing properties. Thus chlorophyll molecule by a specific protein complex containing manganese ions is able to receive the missing electron from the molecule of water. As a result this process the hydrogen protons and oxygen, the product called "side" photosynthesis is obtained. The result of electron beam movement are two compounds: ATP and NADPH. Produced so-called power of assimilation is necessary to carry out the dark reactions of photosynthesis. A further step of photosynthesis does not require solar energy. At this stage there is a series of reactions called the Calvin cycle where carbon dioxide is assimilated from the air. Subsequently, carbon dioxide is converted and in support of enzyme Rubisco is build in natural organic molecules [9].

Formation one molecule of glucose ($C_6H_{12}O_6$) required six Calvin rotation cycle and the energy in the form of 18 ATP molecules and 12 NADPH molecules.

A series of reactions, which constitute to the process of photosynthesis, are initiated by described light beam, falling into assimilation plants dye. Falling solar rays must have sufficient energy to cause an electron excited state in chlorophyll.

For full execution of photosynthesis, and specifically to reduce 1 mole of CO_2 molecule is assumed that it takes 8 photons of red light having a wavelength ($\lambda = 680nm$), as shown in reaction 10.

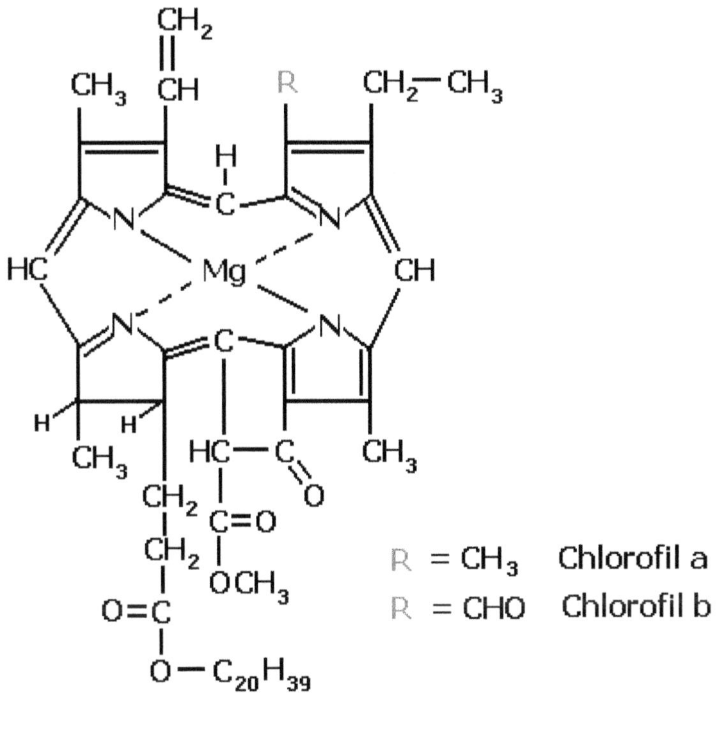

FIGURE 6: Differences between the structure of chlorophyll a and b [6]

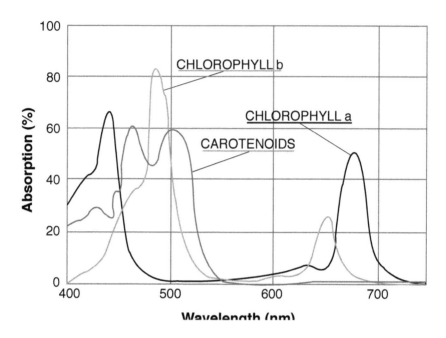

FIGURE 7: Absorption spectrum of carotenoids and chlorophylls a and b is a graph of light absorption by the different wavelengths of light [10]

$$2H_2O + CO_2 + 8\ hv \rightarrow CH_2O + O_2 + H_2O;$$
$$\lambda=680nm = 1.81eV/hv= 174kJ/mol; \Delta G^\circ=479,1kJ/mol \tag{10}$$

One photon of this wavelength has energy = 174kJ/mol, therefore energy used in reaction is: 8 x 174kJ/mol. The free energy (ΔG°) of CO_2 reduction reaction to CH_2O, amounts 479.1 kJ/mol, than photosynthesis efficiency ηf can be calculated:

Efficiency of photosynthesis (η_f) = Energy used/Energy Supplied;

substituting: η_f = 479,1 kJ mol^{-1}/8 174,0 kJ mol^{-1} = 0,34 kJ mol^{-1}

Thus, the ratio between the used energy to the energy put into the photosynthesis process amounts approximately 34%. However, in real conditions only this part of the radiation that can be absorbed by the Earth's surface should be considered. However only about 0.1% of this energy takes part in photosynthesis. In more precise calculations and considering such losses such as "photobreathing" of plants or microbiological decomposition, in practice the efficiency of photosynthesis does not exceed 5% [11].

Each year around the globe are synthesized billions tones of organic compounds that serve as nutrients not only for producers, but also for living organisms, which are on higher levels of ecological pyramid. Biomass as an organic matter is simply result of change-over driven by solar energy.

9.4 BIOMASS AS AN ENERGY SOURCE

Solar energy, as a result of guided for over 3.5 bn years of photosynthesis, is accumulated in the form of organic matter called biomass. In this way, solar radiation is converted into a solid form, which can be used further, not only by organisms in the food chain, but also can provide a raw material for the production of effective energy.

The definition of biomass according with Directive 2009/28/WE is: "biodegradable fraction of products, waste and residues from biological origin from agriculture (including vegetal and animal substances), forestry and related industries including fisheries and aquaculture, as well as the biodegradable fraction of industrial and municipal waste"[12].

However, this definition does not fully covers the meaning of biomass, because its forms are not just a production side effect of other converting processes. There are also dedicated plantation for energetic purposes only. In other words are established the specific energy crops cultivation.

Biomass energy resources are divided into three main groups: [13,14]:

- agricultural biomass (energy crops),
- forest biomass (firewood, waste from wood industry and paper industry),
- and all organic waste from agriculture, forestry as well as gardening.

The basic division of biomass and their byproducts are distinguished as follows due to its physical state:

- solid biomass (that is wood, straw, energy crops, briquette, pellets),
- liquid (liquid and gaseous) energy carriers from biomass processing,

Depending on the degree of physical, chemical or biochemical processing of biomass, it can create:

- primary energy sources—wood, straw, energy crops,
- secondary energy sources—slurry, manure, sewage sludge and other organic waste,
- processed energy sources—energy carriers, that is biofuels (bioliquids) [14,15]

The potential of biomass in the world according to the German Electricity Industry Association (VDEW—German Verband der Elektrizitaetswirtschaft) is about 150 bn Mg per year, which equal to 120 bn Mg of coal. Resources of this biomass exceeds over ten times the current world demand for energy. However, from this potential biomass amount, only about 20–30% is suitable to be used, and in fact only 6 bn Mg of plant biomass is suitable for use [16]. Biomass resources in Poland are estimated at 30 mln Mg per year, which is the energy equivalent to 16–19 mln Mg of coal/ year and is about (110... 130) TW [16] (an average of about 11% of total consumption). For example, to supply biomass to power plant with a capacity of 1MW (= 0.000001 TW) about 5000 Mg of dry weight of the raw material is needed. This amount is equivalent to the coverage of the field area less than 500 ha, and assuming 10% of the density the crops, which is to cover an area of approximately 50 km^2 [17]. The sample calculations show, that required energy consumption for biofuels from biomass in the EU's cover area about 79 000 km^2, which can be compared with the area of the Czech Republic [18].

According to the above figures, the existing biomass resources, and thus the energy efficiency of this raw material is not enough to cover the world's growing demand for energy. It is mean the agricultural land used for food crops should be replaced to a large extent by energy crops plantations, and this way increase the food prices.

Biomass as a solid biofuel is a raw material after suitable processing is used for heat and electricity generation. Conversion of solid biomass for energy purposes can occur by two main pathways (Table 1 and 2) [14, 7]:

1. First one include processes of direct combustion of solid biomass such as in boilers and power plants or by biomass gasification and then combustion of the resulting gas (syngas).
2. The alternative path of energy accumulated in the biomass is to process biomass to liquid biofuels, which are liquid energy carriers for transport.

TABLE 1: Technologies of the direct use of solid biomass (pathway 1)

Process	Combustion	Gasification
Conditions	The excess oxygen in the combustion chamber Temperature: 800-1450°C	Phase I: drying and degassing of the material the shortage of oxygen, temperature: 450-800 ° C Phase II: combustion gases in the presence of excess oxygen, at: 1000-1200 ° C Phase III: heat transfer in the heat exchanger
Final products	CO_2, H_2O	Mixture of gases: CO2, N2, H2, CO, CxHy, CH4 and others.
Energy	Heat energy/ electricity	Heat energy/ electricity
Effectivity	~15-20%	~35%

Combustion of solid biofuels is the most common and easiest, but inefficient method of getting energy. Because the biomass is a mixture of organic compounds such as carbohydrates, the combustion reaction can be presented in simpler form of equation (11):

$$[CH_2O] + O_2 \rightarrow CO_2 + H_2O \qquad \Delta H = -440 \text{ kJ} \qquad (11)$$

Above equation is a simple reverse of the photosynthesis reaction. It can be assumed that this form of solar energy conversion, accumulated in the form of a green plant organic matter is completely depends on carbon dioxide emitted to the atmosphere. As it has been calculated, during the combustion of fossil carbon for every 1 GJ of energy is produced 112 kg of carbon dioxide emissions. In this calculation for biomass (assuming the simplified formula—CH_2O) the amount of CO_2 emitted per each 1 GJ of energy is around 100 kg, which is slightly advantage over the coal. However, it is suggested, the final energy balance of biomass combustion process is zero, because of CO_2 emitted during combustion process is equal to the amount of gas which is absorbed by the organism during the plant growing season [4]. However, the rate of biomass growing in time is not enough, to compensate the CO_2 emitted into the atmosphere. The plants during the growing season, green organisms need about 10 years [4] in ideal conditions for example with proper lighting, temperature and humidity. In order to support the performance of biomass energy crops increase are also used different mixes, fertilizers and pesticides, which are produced from natural gas. It is estimated, to get a 25% share of biomass fuels will lead to increase use of NPK fertilizers by 40% [4, 19]. Plant protection products and supporting substances causes additional emissions of greenhouse gases into the atmosphere such as N_2O. Because of combustion of the solid biomass is not only CO_2 or N_2O emission, but also carbon oxide (CO), sulfur compounds, or other combustion residues such as fly ash and dust.

The efficient combustion of such material needs advanced techniques for boilers, allowing the complete combustion of volatile products of biomass pyrolysis. The incomplete combustion of biomass significantly reduces the efficiency of the process. In addition, low-density of biomass causes difficulties in its storage and transport, which creates extra costs. A large range of moisture content, within the ranges 50... 60% gives the energy value (6... 9) MJ/ kg. On the other side biomass dried has calorific value up to 19MJ/kg [20]) which is also difficult and costly for preparation to use in the heat and electricity production. Consequently, because of the above reasons, considering the whole process, namely from the operation of farm machinery, through the transport, processing of biomass (drying, degassing of material, seal or change of boilers). Finally the combustion of biomass is not efficient to cover the still increasing demand for energy. A good solution of the shortage problem of biomass used for energy purposes could be manufacturing biomass by microorganisms such as algae, which will be discussed later in this chapter.

The mentioned alternative process for the recovery of solar energy collected in photosynthetic organisms is subjected to physical, chemical or biochemical conversion of biomass (pyrolysis, gasification, fermentation, distillation) [4]. In this way the converted form of biomass is called liquid biofuels, which sample preparation processes are shown in Table 2.

9.5 BIOFUELS

In 1900 at the World Exhibition in Paris, Rudolf Diesel introduced the high-pressure engine driven with clean oil from arachidic peanuts. Twenty years later Henry Ford introduced the possibility to power the internal combustion engines with spark ignition by ethanol. Mixtures of alcohol and gasoline as a fuel (30:70 v/v) was reported for using in 1929 in Poland. One of the first scientific publications regarding the possibility of using these compounds to power the engines of tractors has been published in Polish journal "Chemical Industry" [22].

In the expected terms of production and use of biofuels (alternative fuels), it is assumed that those fuels should:

- be available in sufficiently large quantities;
- demonstrate a technical and energetic properties of determining their suitability to supply the engines or heating devices;
- be cheap in production with an attractive price for custumers;
- have a lower risk for environmental degradation than fossil fuels, with less emission of toxic compounds and greenhouse gases in the combustion process;
- provide an acceptable economic indicator of engines or boilers and safety, and enable the lower operating costs of the equipment;
- increase the energy independence.

Until now, there are several complementary definitions of biomass. The previously cited European definition mentioned in the European Directive 2009/28/EC defines biomass as the main source of raw materials for the biofuels production. The European definition distinguishes two basic raw material pathways and corresponding technologies, namely BtL processes ("biomass to liquid"), alternatively BtG ("biomass to gas") and WtL ("waste to liquid"), alternatively WtG ("waste to gas").

The Directive also introduced bioliquids as liquid biofuels for purposes other than transport, like production of electricity, heat and cooling. The introduction of "xtL processes" identifies general processes for conversion of biomass instead of distinguishing of second, third and fourth generation biofuels. The term "synthetic biofuels" is also defined, describing them as synthetic hydrocarbons or their mixtures gained from biomass, such as SynGas produced in the gasification process of forest biomass or SynDiesel.

This division resulted from the conditions described above, and first from the assess the suitability of fuels in modern engine technology and availability of raw materials and their impact on the environment. The formal division of biofuels on the appropriate generations has been published in the report "Biofuels in the European Vision, the Vision 2030 and Beyond." This report divided biofuels into the first generation biofuels, so-called "conventional biofuels" and the second generation biofuels, so-called "future biofuels."

First generation biofuels ("conventional biofuels") include:

- bioethanol (BioEtOH, BioEt), understood as a conventional ethanol gained from hydrolysis and fermentation processes from raw materials such as grain, sugar beets and so on.
- pure vegetable oils (PVO-pure vegetable oils), from cold pressing processes and the extraction of oilseeds;
- biodiesel, which is rapeseed oil methyl ester (RME) or methyl and ethyl esters (FAME and FAEE) of higher fatty acids or from the other oleaginous plants, obtained by the process of cold pressing, extraction and transesterification;
- biodiesel, which is the methyl and ethyl esters gained by transesterification of waste oils;
- biogas got by purification processes of damp waste or agriculture raw biogas;
- bio-ETBE, got from the chemical transformation of bioethanol.

The biofuels in the second category ("future biofuels") have been classified as follows:

- bioethanol, biobutanol and mixtures of higher alcohols and their derivatives got by the advanced hydrolysis and fermentation process of lignocellulose biomass (except of raw materials for food purposes);

- synthetic biofuels, which are products of biomass change-over by gasification and follow by catalytic synthesis for hydrocarbon fuel components in BtL processes.
- fuel for diesel engines gained from lignocellulosic biomass processing through the Fischer-Tropsch processes;
- biomethanol obtained by the processes of lignocellulose transformation, including the Fischer-Tropsch synthesis and also with use of waste carbon dioxide;
- biodimethylether (bioDME) obtained in thermochemical biomass conversion processes, including methanol, biogas and synthesis gases which are derivatives of biomass conversion processes;
- biodiesel as a biofuel or as a biocomponent for diesel engines derived by hydrogen refining (hydrogenation) of vegetable oils and animal fats;
- biodimethylfuran (bioDMF) derived from the processing of sugars, including cellulose, in biochemical and thermochemical processes;
- biogas as a synthetically derived natural gas—biomethane (SNG), obtained through the lignocellulose gasification processes and the appropriate synthesis and by purification processes of biogas from agriculture, landfills and sewage sludge;
- biohydrogen obtained by gasification of lignocellulose and synthesis of gasification products or through the biochemical processes.

The concept of the second generation biofuels is based on the assumption, the raw material for its production should be biomass as well as waste vegetable oils and animal fats, and any residual organic materials, unsuitable for the food industry or forestry.

The Department of Transport and Energy of the European Commission proposed to separate third generation biofuels into those for which the technological development and their implementation into operation can be estimated for the years 2030 and beyond. Biohydrogen and biomethanol were included in this group.

"Fourth generation biofuels" has different definitions, but one of the simplest defines them as crops that are genetically engineered to consume more CO$_2$ from the atmosphere than they'll produce during combustion later as a fuel. Both of these biofuels groups are included in the future biofuels group ("advanced biofuels"). Thus, the third generation biofuels can be got through the similar method as the second generation biofuels, but from the modified (at the cultivation stage) raw material (biomass) by means of molecular biological techniques. The purpose of these modifications is to improve the process of converting biomass to biofuels (biohydrogen, biomethanol, biobutanol), for example by growing trees with low lignin content, development of crops with respectively built enzymes, etc.

The proposal to split the new, fourth generation of biofuels was established due to the need to close the balance of carbon dioxide or eliminate its impact on the environment. Road Map 2050 prepared by the European Commission talks about CCS (Carbon Capture and Storage) [23], but repeatedly claimed that it is a commercial unimplemented technology. Many companies and organizations have social and environmental objections to implement it. Something that is more reasonable and what should be lean is CCU (Carbon Capture and Utilization). It would be profitable to capture CO_2 from the atmosphere or the exhaust of power stations and convert it to fuel by using a sustainable source of energy such as sunlight. Photocatalytic or termochemical conversion of CO_2 to fuels by using semiconductors and metal oxides are two main routes. There is also combination of these routes by using water or hydrogen co-fed with CO_2 for fuels generation.

There are developed new, forward-looking technologies in the U.S. and Europe characterized by high reduction of CO_2 by means of LCA (Life Cycle Assessment) parameter of specific biofuel like:

- biofuel production technologies, including Jet type, by culturing algae without sunlight from agricultural sludge, grass and waste substances, with the use of carbon dioxide (technology "SOLAZYME");
- technology of plasma gasification of waste biomass and municipal and industrial waste (BtG and WtG processes), followed by processing upgraded synthetic gas towards liquid biofuels like GTL diesel and Jet type fuel ("SOLENA" technology, carried out in the UK and Italy);
- technology of carbon dioxide use in the carrier energy production processes;
- complex biorefinery technologies.

The European strategy mentioned in the "European Strategic Research Agenda, Update 2010" defines the following biofuels as a prospective and showing the technological pathways of its getting:

- synthetic fuels/hydrocarbons from biomass gasification (application: transport fuels from RES for jet and diesel engines);
- biomethane and other gas fuels from biomass gasification (substitutes of natural gas and other gas fuels), (application: engine fuels and high efficient energy production);

- biofuels (bioliquids) from biomass got through the other thermochemical processes such as pyrolysis (application: heating fuels, electricity production or indirectly through the xTL processes for transport fuels);
- ethanol and higher alcohols from sugars containing biomass (application: transport fuels from RES or as a petrol biocomponents, E85);
- hydrocarbons from biomass, got from sugars, created in biological or chemical processes (application: renewable transport fuels for jet and diesel engines);
- biofuels gained from the use of carbon dioxide for the microorganisms production or from the direct synthesis of carbon dioxide in thermal and biochemical processes (applications: transport fuels and for aviation).

It is necessary to determine the best and universal technologies of their production. The technologies of biofuels production have to provide the possibility of safe operation of engines while reducing exhaust emissions of toxic ingredients under the needs of the relevant class "EURO" in Europe and the U.S./ California ULEV in the USA. At the same time it is important that biofuels with a similar composition and properties should be available in every country, because of the motor requirements.

That's why the developed by the International Energy Agency, road map for biofuels in transport ("Technology Road map. Biofuels for Trans-

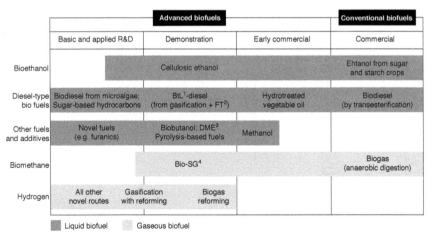

1. Biomass-to-liquids; 2. Fisher-Tropsh; 3. Dimethylether, 4. Bio-synthetic gas.

Source: Modified from Bauen *et al.,* 2009.

FIGURE 8: Biofuels division and stage of its production (Technology Roadmap Biofuels for Transport © OECD/IEA, 2011) [24] with the consent of: OECD/IEA.

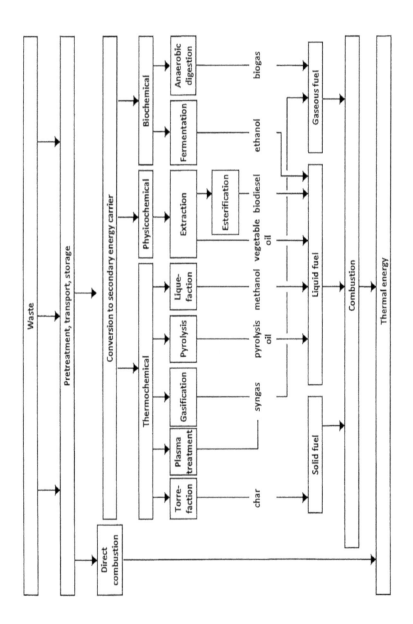

FIGURE 9: Waste to Liquid and Waste to Energy conversion technologies [25], with the consent of: Anouk Bosmans

port") determines the perspective biofuel technologies in the world till 2050. Therefore it was proposed to divide the conventional ("conventional") and future ("advanced") biofuels. Distribution and severity of various biofuels in both groups, according to the MAE, is shown in Figure 8.

The target biofuels production in 2050 required a biomass material with a total energy potential estimated at 65 EJ, which is equivalent to 100 mln ha of cultivations in 2050. Assuming that 50% of the feedstock for the advanced biofuels production will come from waste substances (XtL processes). This means that it is necessary to increase the area devoted for cultivation of biomass for energy purposes. In an optimistic scenario, it is assumed the possibility to get 145 EJ per year of energy from biomass and all kinds of waste substances intended for liquid energy carriers for transport and for heat and electricity in the polygeneration. Estimated reducing of CO_2 emissions is determined at 2.1 Pg per year, with the share of biofuels of 27% (v/v) of the total amount of transport fuels.

Considering the demand for biofuels that meet the future needs for transport, including air transport, and their ability for CO_2 reduction, it was established by MAE the following biofuels and their technology paths:

- fuels from BtL processes (synthetic hydrocarbons compositions), get by fast pyrolysis, by heating biomass at temperature between (400... 600) ° C, and then applied fast cooling. The volatile compounds may be converted to a bioliquid or further catalytic deoxygenation, distillation and refining toward fuel component. The solid residue is called "Bio-char" ("charcoal") and as a by-product can be used as a solid fuel, or used as a sequestration agent and for soil fertility;
- diesel oil from BtL processes, so-called FT-diesel obtained by conversion to the synthesis gas and catalytic Fischer-Tropsch synthesis (FT) in a wide range of liquid hydrocarbons, including synthetic diesel and JET type biofuels;
- hydrorafinated vegetable oil (HVO) as a fuel for diesel engines or heating oil produced by the hydrogenation of vegetable oils or animal fats (non-food and waste). The first commercial plants started in Finland and Singapore;
- cellulosic bioethanol from lignocellulosic raw materials produced by the biochemical conversion of cellulose and hemicelluloses followed by fermentation of sugars (IEA, 2008a);
- biogas obtained by anaerobic digestion of raw materials, such as organic waste, animal manure and sewage sludge, than purified to the biomethane form (SNG!!) by removing CO_2 and hydrogen sulfide (H_2S). Can be applied in transport as methane fuels (bioCNG, bioLNG) or as a source of hydrogen, also for the fuel cell;

- dimethyl ether (bioDME) as gas fuel for diesel engines, derived from methanol catalytic dehydration process, from synthesis gas by gasification of lignocellulose and other biomass. BioDME production from biomass gasification is at the demo phase (September 2010 in Sweden (Chemrec));
- biobutanol have a higher energy density than ethanol in petrol. It can be shared through the existing network for gasoline, Biobutanol can be produced by sugars fermentation using bacteria, *Clostridium acetobutylicum*. Demonstration plants are working in Germany and in the USA, the others are under construction.
- furan fuels in the latest "Technology road map, biofuels for transport," compiled by the International Energy Agency, were classified as prospective biofuels. Compounds such as furfural and 5-hydroxymethylfurfural (HMF) can be obtained with good yield through dehydration of monosaccharides, such as hexoses (e.g. fructose) or pentoses (e.g. xylose) in the presence of various catalysts. Fuel is suitable for spark-ignition engines with the advantages when compering with ethanol.
- solar fuels, obtained by gasification of biomass towards synthesis gas by using heat produced by concentrating solar energy. They can also be obtained by the evaporation of water (water vapor) and the use of carbon dioxide to form synthesis gas catalytically transformed into fuel fractions. In terms of the production of these fuels can be included technology of so-called "Artificial leaf;"
- biorefinery for liquid fuels production and chemical by-products, which discussion exceeds the scope of this chapter.

In terms of the most promising biomass for the future biofuels production, considering so-called "land hunger" ("ground competition") and the needs in terms of CO_2 emissions, can be distinguished following cultivation: algae, camelina, jatropha, and halophytes. Under development are new technologies without sunlight ("dark"), marine photosynthetic membrane systems for the algae production, as well the technologies for biomethanol production.

9.6 TECHNOLOGIES OF LIQUID ENERGY CARRIERS PRODUCTION BASED ON WASTE SUBSTANCES

9.6.1 WTL (WASTE TO LIQUID) AND WTE (WASTE TO ENERGY) PROCESSES

Dynamically observed population increasing in the world stimulates fast economic growth. This is related with growing demand for all kinds of products and energy. In result of progressive consumption, waste genera-tion is increasing. According to the data total amount of waste, which is generated in the European Union, maintains an upward trend. Municipal solid waste (MSW) amounted 150 million Mg in 1980. In 2005, increased up to 250 million Mg, and forecasts for 2015 growth is estimated up to 300 million Mg [25]. The constantly increasing amount of waste contributed in overfilled existed landfills, forces society for development of new land-fills. The overfilled landfills are big the social and environmental problem. Therefore, novel energy technologies have preferences for waste treatment WTE (Waste to Energy), or for receiving new products from waste in the process of WTP (Waste to Product.). Waste treatment processes for liquid energy carriers are processes WTL (Waste to Liquid). The trends for con-ventional energetic waste utilization are based on direct combustion and more advanced thermochemical methods such as pyrolysis, gasification, and plasma technologies under development since 1970. Figure 9 shows the main alternative methods of converting waste to energy in WTL and WTE processes [25].

In the WTL and WTE processes, pre-treated waste is converted into secondary energy carriers using thermochemical methods: physico-chem-ical or biochemical. As a result of these changes is obtained biofuels solid, liquid or gas. These fuels are burned and converted into thermal energy. WTL processes path is shown in Fig.9..

TABLE 3: Basic physical and chemical conditions of the process HTU [26]

Conditions	Temperature: (300 ...350) °C
	Pressure: (120...180) bar
	Reaction time : (5... 20) minutes.
Feedstock	All types of biomass, domestic, agricultural and industrial residues, wood.
	Also wet feedstocks, no drying required.
Chemistry	Oxygen removed as Carbon Dioxide
Products	45 Biocrude (%w on feedstock, dry basis)
	25 Gas ("/> 90% CO2)
	20 H$_2$O
	10 dissolved organics (e.g., acetic acid, methanol)
Thermal efficiency	70–90 %

A HTU is processing waste into liquid biocrude oil (Hydro-Thermal Upgrading), at high temperature and pressure near supercritical condition of water, which is usually the solvent in this process, the basic conditions are shown in Table 3.

HTU process developed by Shell in 80's occurs at a temperature up to 350°C and a pressure about 180 bars. HTU can be used to convert liquid fuel from a wide range of biomass feedstock, without the need of drying raw material. It has been designed to carry out the reaction in the excess of water, under supercritical condition of water. The final product is a "bio-oil" ("biocrude") with properties similar to crude oil, so it can be used after upgrading as fuel in boilers, turbines and so on. The calorific value (LHV) of biocrude obtained this way amount 30... 35 MJ/ kg [27].

Another an example of a technology for waste conversion is based on the catalytic thermal depolymerization patented by Alphakat GmbH, which principle of operation is shown in Figure 10.

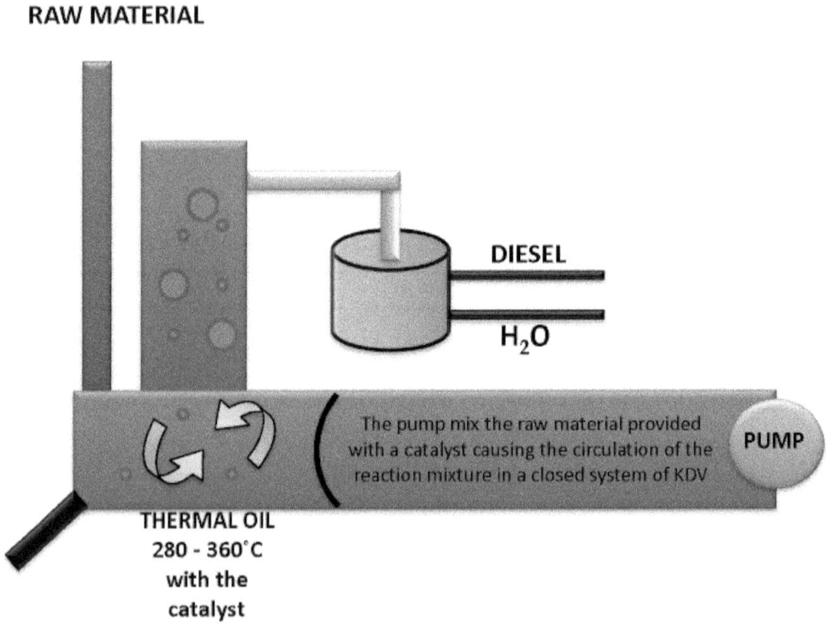

RAW MATERIAL

DIESEL

H$_2$O

The pump mix the raw material provided with a catalyst causing the circulation of the reaction mixture in a closed system of KDV

PUMP

THERMAL OIL 280 - 360°C with the catalyst

FIGURE 10: The principle operation of KDV technology [28 modified]

KDV process takes place in a special industrial installation known as KDV Unit. As can be see from Figure 10, the whole system is a closed with strictly defined conditions: the process is carried out at ~350°C, under low vacuum 0.9 bar. The vacuum is maintained in the system by special vacuum pumps. As shown in the figure, the pump mix the raw material provided with a catalyst causing the circulation of the reaction mixture in a closed system. After heating the batch by thermal oil up to ~350 °C, volatile organic compounds (VOC) are formed, which at the main distillation column undergo separation for diesel and gasoline fraction [29, 30].

Using KDV technology a wide range of raw materials derived from biomass or waste materials, either organic or mineral, can be process. Table 3 summarizes the basic substances that can be used as a feedstock in KDV plant. We can see all kinds of biomass waste, agricultural, municipal and industrial sewage, as well as synthetic materials.

TABLE 4: Basic substances which can be used as raw material for the KDV installation [29, 30]

Lp.	Type of material	Sort of material
1	Biomass (C-3 and C-4 plants)	· grown energy sources like Jatropha,
		· wood, biogenous residues like leaves, straw, etc.
2	Waste	· Industrial Waste (IW),
		· Municipal Solid Waste (MSW),
		· agricultural waste,
		· waste oil (also contaminated oil),
		· refinery residues, bitumen. etc,
		· dried sludge from sewage treatment plants,
		· rubber and tires.
3	Synthetic materials	· All kinds of plastics and synthetic materials (PVC, PP, PET, etc.)

The main products from KDV plant is synthetic diesel fuel or kerosene depends on process parameters [30]. However these fuels are available after catalytic hydrotreatment process (like hydrodesulphorization in refinery), because the product after distillation has still properties of biocrude oil. KDV were started as a demo units in Germany, Canada, Spain and Mexico and recently in Poland [31].

9.6.2 OVERVIEW OF SELECTED TECHNOLOGIES FOR LIQUID BIOFUELS FROM WASTE CARBON DIOXIDE

9.6.2.1 CONVERSION PROCESSES

Scientists all over the world are trying to learn the essence of the photosynthesis process. Already in 1912, the Italian chemist Luigi Giacomo Ciamician in the Science magazine, said the society should be transfered from a civilization based on oil and coal and start using clean energy from the sun. He thought that by using appropriate photochemical reactions and using the new compounds is possible to discover the mystery of the photosynthesis guarded plants.

In July 2010, in the United States was established Joint Center for Artificial Photosynthesis—JCAP, funded by the U.S. Department of Energy (DOE). The aim of this project is to develop a method able to copy the photosynthesis process, available to establish the path for the production of liquid fuels using waste carbon dioxide.

According to information provided by JCAP [32], major challenges facing scientists dealing with the artificial photosynthesis is a system that consists of key elements of the process: the visible light absorber, catalyst and membrane switches. Visible light absorber is a molecule that is designed to capture and convert the sunlight with the high efficiency, like the chlorophyll in plants. Absorbed energy puts absorber molecules into excited state, in which all the light energy is submitted to electron. Properly designed absorber pushes the electron from the shell, directing him to an adjacent molecule (in the same way as it does in natural photosynthesis, where electrons jump from photosystem I to photosystem II). In this way formed electron beam passes through a further conveyor releasing the energy required for further electrochemical reaction. At the point of detachment of the electron take place creation so-called. "Hole," or positive charge. At this point is necessary a catalyst that efficiently breaks the water molecule, this way is freed the missing electron in place of created "hole." In natural photosynthesis complex of the protein, which contains manganese ions act as the catalyst. This compound is still a great mystery because it is not yet possible to copy of its complex structure in the laboratory. Therefore, research is continuing to find a suitable replacement, which will perform a similar sequence cleavage involving water molecules processed by the solar energy absorber. Catalyst must not only be efficient and sustainable, but also cheap, and widely available on the Earth, preferably a non-Noble metal or a complex.

Mimicking plants, we can create something like artificial leaf. The function of the matrix is not only to act as a micro bioreactor, but also play role of separation reactants and products (fuel and oxygen). Oxygen is not a desirable product in artificial photosynthesis. Instead of molecular oxygen (O_2), it can happen that forms oxygen radicals (O), very reactive with strong oxidizing properties.

With a listed part of the process of artificial photosynthesis, there is only lack of one element—the connector, through which various components interact with each other and create a functional whole. In order to build a system composed of the described elements, there are carried out intensive research in the USA, Switzerland, UK, Japan and Poland.

The project regarding artificial photosynthesis in Poland started in 2006 at the University of Maria Curie Skłodowska (UMCS) University in Lublin and in collaboration with EKOBENZ company. The concept of methanol synthesis based on the carbon dioxide formed as a result of different technological processes has been developed by Professor Nazimek from UMCS. The invention is based on the oxidation of water using titanium dioxide (TiO_2) as a photocatalyst and at the same time the reduction of carbon dioxide (dissolved in NaOHaq). Reaction is carrying out in assistance of fixed ultraviolet radiation (UV) as is shown by reaction 12.

$$(2CO_2 + 4H_2O)hv \rightarrow 2CH_3\text{-}OH + 3O_2 \tag{12}$$

This technology will enable the production of methanol from the two cheap, universal components: carbon dioxide and water vapour. The proposed technology is one of the most promising methods of utilization of CO_2, since this compound can be converted to a valuable product, under natural conditions by exposure to UV light.

Implemented technology is characterized by the synthesis reaction of carbon dioxide and water which takes place in aquatic environment at the presence of a photocatalyst at a temperature of 20 ° C and atmospheric pressure. Specifications referred this reaction on a laboratory are shown below:

- Energy consumption 0.75 kWh/dm³ of methanol
- the solution flow rate 8 dm³ / h;
- carbon dioxide flow 370 dm³/ h;
- the concentration of methanol in the aqueous solution was 15%.

Based on these parameters has been developed preliminary schematic diagram of the planned production process, as shown in Figure 11.

Under the proposed scheme carbon dioxide is stored in the reservoir of raw material from which goes to the mixer where is absorbed in demineralized water. Then, in the reactor under the influence of UV radiation takes place photoreduction reaction. When reaction is completed, the distillation take place and methanol (CH_3OH) and water (H_2O) are separated. The methanol is sent to the product tank, and the water is reused in the production process—is transfered to the mixer. The active and stable catalyst is the key for obtaining methanol by CO_2 photoreduction. For photocatalytic tests were applied TiO_2 in the form of anatase supported by alumina as a carrier of the active phase.

The study at laboratory scale was carried out in flowing system, at atmospheric pressure and ambient temperature, equipped in quartz pipe reactor. In typical experiment 100 cm³ of distilled saturated by CO_2, and air or nitrogen flowed through reactor. Bed catalyst was exposed on UV lamp radiation at fixed wavelength. The effluent was analyzed for content of organic compounds by using gas chromatograph. The conversion of CO_2 to methanol was determined by the formula (13).

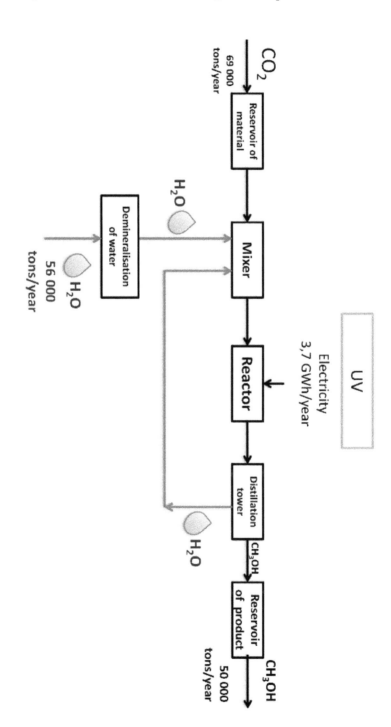

FIGURE 11: Block diagram showing a manufacturing process of methanol carbon dioxide and water, with the consent of: Ekobenz Ltd

$$X\% = C_0 - C_1 / C_0 \tag{13}$$

where: C_0: initial concentration CO_2 [mol/dm^3] and C_1: final concentration CO_2 [mol/dm^3]

The conversion degree of carbon dioxide amounted only 6% [33]. It was found the titanium dioxide catalyst had a defect in the absorption spectrum, which was located only in the UV light range. Visible radiation even during long-term irradiation was not able to induce the active phase of the pure TiO_2. The suspension of titanium dioxide could be the solving the problem, but it is related with separatation of the catalyst from the product liquid phase after reaction completion.

Therefore, the above measurements were carrying out involving a new nano-structured wall catalyst. This catalyst is protected by Polish national patent No. 208030. The catalyst is characterized as the active sites in the form of clusters TiO_3 with aluminum III ions and sodium I ions (TiO_3 content in the range 4.36 to 5.34% (m/m) of the total weight of the catalyst) supported by alumina. These clusters are deposited on the inner wall of the aluminum tube of the reactor.

The catalyst described above as the wall form allows to implement an artificial photosynthesis process in the new technology. Developed a new efficient photocatalyst created the conditions for process with performance sufficient for commercial applications, for example by design 0.5 m length of processing pipe reactor, photocatalysts (with $\varphi = 5$ cm) allowing to convert the 370 dm^3 of CO_2/h into methanol with the degree of conversion at 97%.

Figure 12 shows the schematic diagram of the installation for methanol production applying artificial photosyntesis.

As shown in Figure 12 carbon dioxide with water are transfered into mixers arranged in series, which supply the reactors system. The catalyst is located on the reactor walls, as a result of reaction are products like oxygen molecule, water and methanol. Molecular oxygen is discharged to the gas collector, and the mixture of methanol and water is further transferred into the distillation tower, where separation these compounds take place. Water is returned to the mixer, where it is used in the subsequent cycles of the process, and desired the final product is separated as a pure methanol.

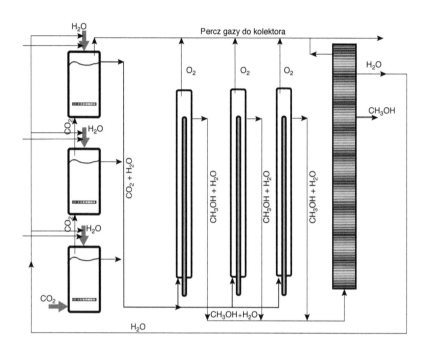

FIGURE 12: A schematic diagram of the measuring system for the production of methanol in the artificial photosynthesis, with the consent of: Ekobenz Ltd

The methanol obtained through the described technology can be directly applied as a fuel in dedicated engines and adjusted fueling systems with seals, proper polymer pipes because methanol is known as a good solving agent for many materials used in cars now. On the other hand methanol can be catalytically converted towards gasoline fraction by using very well known from 70's Mobile process—MtG (Methanol to Gasoline). MtG method involves the synthesis of a higher hydrocarbons mixture (which are synthetic petrol) from methanol, the intermediate product according to the general reaction (14) are:

$$nCH_3OH \text{ (zeolite catalyst eg. ZSM-5)} \rightarrow CnHn + x + H_2O* \qquad (14)$$

*The n and x values are variable depending on the temperature and pressure in the system, as well as catalyst used in the process.

Reaction (14) occurs at a temperature $T = 723K$ and under the pressure of $p = 200$ bar. Water is a by-product of the reaction is, which is a limiting factor in MtG synthesis because it affect the catalyst performance. The catalysts used in the synthesis of synthetic gasoline from methanol are based on aluminosilicate matrix. These type of catalysts in some excess of water vapour, in the reaction mixture, lose their activity because of extraction for example Al ions from the matrix of catalyst. A new catalyst has been developed, which is water-resistant in the reaction mixture used in MtG method as well as related technology EtG (ethanol to gasoline, which will be discussed later in this chapter). Nano-structure catalyst developed for the synthesis of gasoline from the ethanol according to equation (14) is characterized by the active centers in the form of a copper ions coordinated in octahedral structure. Copper is supported by nano-structured aluminosilicate matrix with metal content in the range of 0.2-0.5% w/w. This catalyst has a high water resistance with content even more than 10% water by volume of the reaction mixture.

The final product of the described method MtG is appropriate mix of higher hydrocarbons, which are synthetic gasoline with an octane rating of up to 108, characterized by the same properties as gasoline derived from crude oil processing. Summary of the final products of described MtG method is shown in Table 5 [34].

TABLE 5: Summary of the final products of the MtG technology

Gasoline—until 108 LOB (Q = 44,8 MJ/kg)
ON—56 cetane number
SNG (~ 95% CH4)
aromatic fractions

According to preliminary estimates, this method is able for application within three years, and what is most important can cause the reduction of CO_2 in the atmosphere by 25%.

Some inaccuracies in the calculation of the energy balance above process should be noted. The synthesis of methanol from CO_2 and H_2O is endothermic reaction, which means that for the occuring reaction is required determined amount of energy. It was found out, the total power consumption in order to obtain a 1 dm³ methanol is 0.75 kWh. The balance of energy showed the following conclusions:

- if: 1 kilomole of CH_3OH = 586 MJ
- 1 kilomole of CH_3OH = 32 kg
- therefore: 32 kg CH_3OH = 586 MJ
- Then 1 kg CH_3OH is: 586 MJ / 32 kg = 18.31 = 5.08 kWh
- specific weight CH_3OH is: 0.48 kg/l.
- therefore: 1 kg = 5.08 kWh, which gives 0.48 kg (=1 liter) = 2.44 kWh
- Conclusion: To make 1 liter of methanol from CO_2 and H_2O is required 2.44 kWh, it is three times higher than reported by the inventor of the method— 0.75 kWh.

Therefore, the cost obtaining of 1 liter methanol will be about 0.31 USD, and not as referred value 0.03 USD [35]. Regardless, a drawback of this method is the fact the developing technology is based on the pure components. In the synthesis reaction of the methanol with carbon dioxide and water, is used 96.5% CO_2 from the pressure cylinder, which occurs very rarely, only in some chemical processes (including fermentation). The effects of using carbon dioxide mixed with air in the reaction, such as is found in the nature, wasn't fully investigated yet.

In summary, the method of converting carbon dioxide into methanol at the moment is not applicable, because there is no beneficial economic and energy effects. For these reasons, the project in methanol synthesis in the process of photoreduction of carbon dioxide was suspended. The chance for this interesting project is replacing artificial sources of UV radiation by the natural sunlight. Then the project possibly would be cost and energy-effective. It depends on the active catalyst that would work efficiently in the visible light range.

The part of the project under investigation is still EtG process (Etanol to Gasoline). The coupling process of ethyl alcohol towards of higher hydrocarbons also take place with a suitable zeolite catalyst, at higher temperature and pressure. Replacing ethanol by its homologue methanol causes a significant reduction of pressure required for a reaction, from about 200 bar to 30 bar. Also, the temperature is much lower and amounts 653 K [36]. Reaction of ethanol conversion into higher hydrocarbons is exothermic, and therefore there is a problem with removal of the heat. if combined with endothermic process of methanol synthesis could improve the energy balance of carbon dioxide conversion into methanol. The tubular reactor is cooled at the moment by medium, which at the beginning of the research was air. Now it was changed to a more efficient oil. This method of catalyst cooling is not very effective, because it reduce the temperature only near the outer wall of the reactor. The gradient temperature within radius can reach the temperature up to 300 °C. Then, the catalyst located in the center of the reactor can be affected by sintering process,

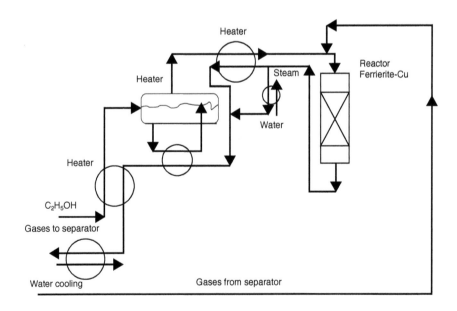

FIGURE 13: Diagram of the research installation for the process EtG, with the consent of: Ekobenz Ltd

and then lose activity. The current research aim is to optimize the temperature of the entire length of the reactor, to avoid deactivation of the catalyst.

At the present stage of research, the experiments are carried out in one reactor, and the target is to operate six reactors cooperating with each other, forming so-called hexagonal reactor. The overall diagram of the EtG is shown on Figure 13.

Under the pictured diagram, ethyl alcohol is sampled into the heater, where in the electric heater spiral is heated to about 270 °C, and through the isolated pipes reaches the reactor. Outlet reaction mixture flows to the condenser, where the liquid products are out dropped. However the ethanol is converted not only towards the liquid fraction, but as a result of this reaction the gaseous products are also formed. Gases are recycled to the reactor, mixed with a new part of ethanol steam and are converted into gasoline fraction. The liquid fuel is further analyzed from point of view requirements of gasoline and diesel. The typical content of such products are benzene (as aromatization compound of ethanol on zeolite), olefins, paraffins and other aromatics. Research in this area is continued in Poland on the frame of "Operational Programe of Innovative Economy". The process is called ETG (ethanol to gasoline) is similar to the MTG (methanol to gasoline) process which was successfully used by Methanex in New Zealand.

9.6.3 PHOTOCONVERSION TECHNOLOGY USING "ARTIFICIAL LEAF"

On 241 Congress of National Meeting of the American Chemical Society in Anaheim, Califfornia, it was announced the invention of the first functional artificial leaf. "A practical artificial leaf has been for decades the one of the Holy Grails of the science"—said the leader of the team, Dr. Daniel Nocera—"We believe that we did this" [37].

A breakthrough moment in the study of artificial leaf is development of a suitable catalyst, nickel (Ni) and cobalt (Co). These elements are widely available comparing with noble metals, what is important they act at ambient conditions and have a high stability during the reaction. Inventor

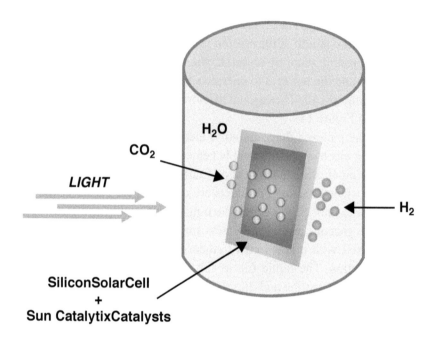

FIGURE 14: Diagram of the artificial leaf [38 modified]

reported very effective work of artificial leave for production hydrogen and oxygen.

The artificial leaf is a matrix composed of material used in most of the solar cells, semiconductor silicon. Surface is covered with a matrix of cobalt catalyst on the one side, and on the other side with the molybdenum-zinc alloy of nickel. Solar radiations falling on the surface are transformed into electricity which allow to decompose bond of water molecules in the presence of catalyst. On the first side of the matrix, the electrons are knocked out from the water molecules using cobalt catalyst. The water molecule is decomposed into hydrogen ions H+ and oxygen. Oxygen remains on first this side of the membrane, while the hydrogen ions are

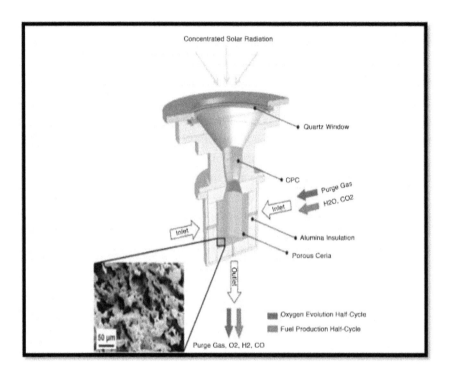

FIGURE 15: Prototype reactor uses the sun to produce storable hydrogen fuel [41 modified]

transported to the other side of the matrix, where the nickel is catalyzing the recombination reaction of hydrogen ions with the previously made free electrons. The result of the process is H_2 molecules formation [37-40]. Diagram of the act "artificial leaf" is shown in Figure 14.

In this way, as it is shown in Figure 14, one surface of an artificial leaf produces oxygen and the other—hydrogen.

According to data the authors team, with one gallon of water, exposed to solar radiation using artificial leaf; it is possible to produce enough electricity required for the daily demand for household energy in developing countries.

9.6.4 SOLAR REACTOR IN TECHNOLOGY OF SYNTHESIS GAS GETTING ("SOLARFUELS")

The research group of CalTech from United States published in December 2010 the paper regarding their latest achievement in artificial photosynthesis. Prototype solar device was constructed, where the heart of device is cerium dioxide (CeO_2) catalyst. Diagram of the device is shown in Figure 15.

Reactor operation is based on the absorption of concentrated solar radiation, which falls on a quartz apparatus window, then passes into the insulated chamber of the receiver, which is filled with a porous cerium oxide. This compound has a specific property of the oxygen binding at low temperatures, and desorption it at high temperatures, without damage of its crystalline structure, according to the reactions (15), (16), and (17).

$$1/n\ MO_2 \rightarrow 1/n\ MO_{2-n} + 1/2\ O_2\ (g)\quad TH \tag{15}$$

$$H_2O(g) + 1/n\ MO_{2-n} \rightarrow 1/n\ MO_2 + H_2(g)\quad TL \tag{16}$$

$$CO_2(g) + 1/n\ MO_{2-n} \rightarrow 1/n\ MO_2 + CO(g)\quad TL \tag{17}$$

where: M—pure cerium or cerium with mixtures; TH—at high temperatures; TL—at low temperatures [42]

In this way, it absorbs oxygen gained from supplied to the reactor carbon dioxide (CO_2) and water vapor (H_2O). Thus, inside the reactor are formed the carbon monoxide (CO) and hydrogen (H_2), which is an ideal source of energy. Hydrogen is considered to be so-called "clean fuel" because in the result of its combustion the water is just produced. However, the reaction of hydrogen with carbon monoxide if presented by stechiometric reaction can be directed towards methane formation, popular heat, transport fuel and source of chemicals production (reaction 18). The synthesis gas in proper molar ratio of CO/H_2 and using FT catalyst (Co or bimetallic with Fe) can be converted towards higher hydrocarbons.

$$CO + 3H_2 \leftrightarrow CH_4 + H_2O \tag{18}$$

In the CalTech process the reactor was exposed to solar radiation at 1.9 kW of power. Such power is widely used in various types of solar antennas or solar towers; any of these devices has not reached the efficiency of the presented prototype. This can be explained by the fact, the reactor uses the entire spectrum of light, not only the selected wavelength. Disadvantages of the presented solar reactor is the energy losses as a result of energy radiation through the walls of the reactor (about 50% of the energy input) and the release of radiation through the quartz window (about 41% of supplied energy).

It can be assumed that further improvement of the process will be able to reduce the temperature of the reactor (today reactor operates at 1648 °C), and less energy losses through more efficient use of solar radiation. An initial analysis of the process effectiveness showed that improvements structure of cerium oxide will allow to increase activity of catalyst in the range of 16-19%.

This concept can be used for conversion carbon dioxide emitted from coal power plant to liquid energy carriers. The pure carbon dioxide is not required to carry out this reaction. This allows at least twice using of the same coal compund [42].

9.6.5 BIOLOGICAL PROCESSES USING CARBON DIOXIDE AS A RAW MATERIAL

As stated previously, the world's biomass resources are not enough for producing renewable energy with at the same time growing needs of modern civilization. For example in UK, replacement of diesel by biodiesel produced from rapeseed would required plantation more than half area of the country [43]. The areas for the production of liquid biofuels is competitive with the same areas for the cultivation of food crops. There is a conflict between energy and food use of agricultural land. In addition, oilseeds cultivation leads to monocultures over large areas, which is inconsistent with the Directive 2009/30/EC. As stated in mentioned Directive production of biomass for energy purposes should go hand in hand with biodiversity. Nowadays more and more we hear about a new raw material for the production of liquid fuels which can be tiny microorganisms such as algae. Algae growing is a good option for limited areas of cultivation of biomass for energy purposes. Algae can be grown in areas not suitable for the cultivation of the soil, such as deserts or oceans. In addition, they are a kind of "factory," able to convert waste carbon dioxide into valuable energy products.

The idea of using algae as a feedstock for bioenergy production has already appeared in the mid-twentieth century. In 1940, it was discovered that many species of microalgae growing in strictly definite conditions can produce large quantities of lipids. The concept of using the stock of lipids as an energy source has been caused by the oil crisis which arose from the imposition of the embargo in 1973 by the OPEC countries. The rapid increase in oil prices and the reduction of the energy carrier from Middle East to the United States contributed to the search for new sources of energy such as micro-algae. In 1978, Program of the Department of Energy for Aquatic organisms have been created (DOE's Aquatic Species Program), which focused on the acquisition of the fuel in the form of pure hydrogen supply by aquatic organisms. Already in the early 80's, this was changed and focused on trying to manufacture liquid fuels, mainly bio-diesel. Over 300 000 existing algae strains were collected from different extreme environments. The next seven years were carried out research regarding tolerance of different salinity on temperature, pH, and the ability for the production of neutral lipids. After these tests, the study was limited up to 300 promising strains, in which the main role was played: diatoms algae (*Bacillariophyceae*), and green algae (*Chlorophyceae*) [44].

Algae are a valuable source of raw material for energy transformation because of fast growth of the biomass. Autotrophic algae contain chlorophyll molecule in their cells so they have the ability to carry out the process of photosynthesis. They are able to use about 10% [45] of the sunlight falling on it, in consequence able to double their weight during the day, and in the experimental conditions this time was reduced up to 3.5 hours [46]. The study shows that wit1hin one year cultivation of algae on the area of one hectare, it is possible to get 8,200 liters of biodiesel from extracted algae lipids, while for other oleic plants like jatropha only 2700 liters, 1560 liters from canola, and from soybean only 544 l [43]. These data are summarized in Table 5.

According to the literature, the production algae biomass is 40... 60 times higher than previously cultivated energy crops. The yield estimated with different cultures growing algae in closed photo bioreactors is about (400... 500) Mg/ ha/ y, which coresponds (400... 500) thousand m3 of bioethanol per year [47]. Another advantage is that algae cultivation is possible in wide variety of water sources: seawater, postproduction water, as well as in wastewater.

TABLE 6: Estimation of oil productivity from different crops [43]

Crop	Biodiesel yield (L/ ha/ year)
Oilseed rape	1560
Soya	544
Jatropha	2700
Chlorella vulgaris	8200

One of the key reasons to use algal biomass for energy purposes is these microorganisms absorb and convert a significant amount of residual carbon dioxide. To provide 100 g of biomass, algae consume about 183 g of carbon dioxide, which is associated with a 50% share of CO_2 in dry weight of these microorganisms [48]. With the ability to bind CO_2 algae opens the possibility of using waste carbon dioxide from the exhausted gase, for example coal-fired plants. In this way, they close the circulation of carbon in nature. They also show the ability to absorb other waste components, such as nitrogen or phosphorus for example from of chemical fertilizers. With this ability algae can lead the process of bioremediation of contaminated environments, thus contributing to reduce contamination of soil or surface waters (eutrophication) and groundwater.

Energy sources that can be produced from algae depend on the substrate used and the method used to process the obtained algae biomass. The main path of bioenergy production using algae is shown in Fig.16.

As shown in Figure 16, we can get bioenergy directly from algal biomass include electricity, syngas, biohydrogen and biomethane. However,

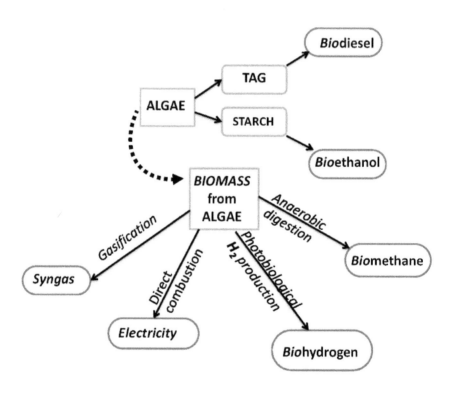

FIGURE 16: Path of bioenergy produced by using algae.

two main liquid energy carriers gained from the cultivation of algae are biodiesel and bioethanol.

For the production of biodiesel from algae, the most important is choosing the right kind of algae for triacylglycerols (TAG) obtaining as well as made the ideal cultivation system of these microorganisms. It is necessary to use not only the knowledge of the bioenginering of the various strains of algae, but also integrate this knowledge with the best matching further process technologies [44].

It was observed that in the optimum conditions for the development of algae, which is a sufficient amount of nutrients (such as nitrogen and phosphorus) biomass increase, but the percentage of the dry weight of TAG is rather small. For example for the species *Chlorella Vulgaris* it is about 14-20% of TAG. However, in the lack of nutrients, algae forms a reserve of energy in the form of so-called TAG. The percentage of TAG per dry mass for *Chlorella Vulgaris* reaches up to 70%, when nutrition substances are limited. One proposed solution for this phenomenon is a two-step process of algae cultivation. It is based on the fact that in the first stage of breeding algae grown in optimal conditions, with a sufficient amount of nutrients, which results in a rapid increase of their dry biomass. In the second stage the algae are transferred to the place with significantly reduced amount of nutrients. So these micro-organisms, as they are stored in a nutrient deficient conditions lead to increased production of TAG cells.

Another solution is to use microtrophic algae. They are characterized that under conditions of sufficient amounts of nutrients, mainly easily assimilable carbon such as glucose, are transferred to the heterotrophic mode during which produces significantly greater amounts of lipids. Unfortunately, by introducing additional nutrients such as glucose, this solution entails the risk of contamination algae culture by heterotrophic fungi and bacteria [49]. Moreover, one of the main advantages of using algae for biodiesel production is the use of carbon dioxide from the combustion of fossil fuels.

Properly grown biomass is collected and processed in such a way to maximize the production of triacylglycerols cells for the production of biodiesel (Fig. 17). Algae farm is usually isolated as was shown in Figure 17. In order to reduce the water content, of the harvested biomass, is added the flocculant. Lipid fraction extraction take place with help of hot diesel fuel, than the mixture is separated by using centrifudge. The oil fraction is separated from the watery phase and the biomass waste. Oil fraction can then be used in transesterification reaction to produce FAME or synthetic hydrocarbons using HDO process. As a result, from showed processes already are produced fuels by Sapphire Energy, Petro-algae and Solix Biofuels.

For example, an American company Sapphire Energy in 2008 has successfully produced 91-octane gasoline from algae, which complies with U.S. standards. They made a flight test with the twin-engine Boeing using algae-based jet fuel in 2009. They started the construction of an integrated minibiorefinery based on biomass from algae in the southern part of the state of New Mexico in 2010 (called Intergrated Algal Bio-Refinery) [50, 51]

In the process of producing ethanol from starch contained in algae cells, the most plentiful in this hydrocarbon algal classes are: Chlorophyceae (green algae) of the types: Chlamydomonas and Chlorella and Cyanophyceae of species: Arthrospira, Oscylatoria and Microcystis. Bioethanol from the sugar gained from algal biomass can be obtained by:

- Breeding of certain species of algae in anaerobic and without the sun conditions (then the algae produce ethanol within the extracellular space);
- Extraction starch from algal cells and then, through the enzymatic hydrolysis and fermentation ethanol is produced [45].

For example, patented by the company Algenol Biofuels process is called Direct to Ethanol ® and is based on breeding of algae Cyanophyta in specially designed flexible plastic photobioreactors. Using seawater and waste carbon dioxide (fed directly from industrial plants) and subjected to direct sunlight, the algae perform photosynthesis. Sugar molecule (starch) get by photosynthesis process is fermented to ethanol, which diffuses through the cell walls of algae to the solution of breeding. Because of raised temperature during the day, ethanol is evaporated from the solution. On the inner wall occurs condensation of ethanol vapors when the temperature decrease at night outside of photobioreactor,. The condensate drops down of the reactor, where on special internal basins is collected and transferred to distillation unit.

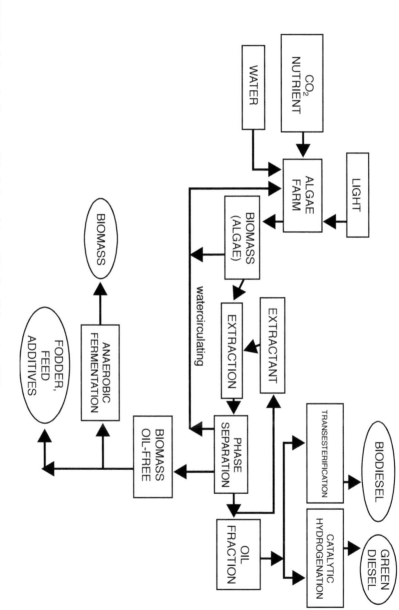

FIGURE 17: Schematic diagram of the fuel receiving a "biodiesel" from algae [45]

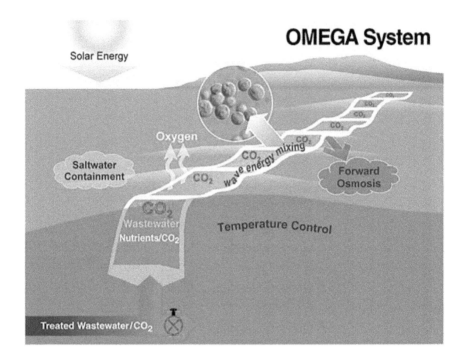

FIGURE 18: The OMEGA System [54], with the consent of: Jonathan Trent

Except lipid molecules in the form of triacylglycerols, or hydrocarbons in the form of starch, which are precursors for the production of biofuels, from algae biomass may be obtained also other products such as proteins. Therefore, in the production of biofuels using algae is formed a wide range of products such as animal feed, fertilizers, pigments, bioplastics, detergents, cosmetics and even food [52].

9.6.5.1 OFFSHORE MEMBRANE ENCLOSURES FOR GROWING ALGAE—OMEGA SYSTEM

OMEGA technology, originally developed by NASA, is derived from the space program aimed to closing the loop (called „close the loop") between the waste stream and the resources necessary for astronauts during long flights [53]. The modified system is to grow cultures of algae in specially designed, floating on the water surface, photobioreactors (PBR) composed of polymers. On the figure 18 is presented the idea of this system.

The system consists of a number of connected, floating photobioreactors, which are pumped by waste water from the mainland (the actual appearance of the plastic photobioreactors—"bags" floating on the water are shown in Figure 19).

Algae consume nutrients contained in the sludge, and they associate the carbon dioxide from the air, or waste CO_2 emitted directly from industrial gas plants. By using solar energy, they embed CO_2 into their cells and give off the oxygen into the atmosphere, at the same time producing biomass and oils in their cells. The temperature inside the photobioreactor is controlled by the heat capacity of the surrounding ocean waters, and the gradient between the wastewater and seawater was used for the drainage of photobioreactors. In case of leak the photobioreactor content into the ocean, freshwater algae are breaking up, because they are not able to survive in saltwater. These biochemical species of algae are decomposed into simpler chemical compounds. Thus sea water where are mounted plantations algae are not contaminated. The algae after the growing phase are transfered into the osmotic chamber to make them more dense. Then biomass material is transported into a collection chamber, from where are send to biorefineries. Waste water is returned to the cultivation unit, to maintain the adequate stock and optimum concentration of nutrients for carry out the photosynthesis process by algae. The OMEGA technology final products are biodiesel, jet fuel and by-products such as cosmetics, fertilizers, and animal feed.

The method of producing biofuels using microorganisms is an agree with the nature. There is not used genetic modifications organism and the risk of GMO species invasion into the natural sea environment. Strains of algae are biodegradable, which means it does not involve any risk of pollution of the seas, and oceans, and may even provide food for fish living there. The only added cost is the pumping wastewater into the photobioreactor. However, it was showed the idea of driving the process using wind turbines or solar panels [54, 55].

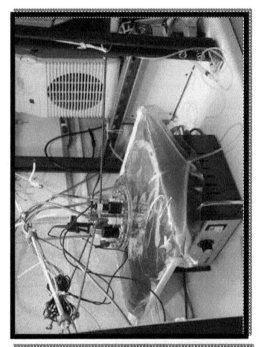

FIGURE 19: Examples of prototype photobioreactors for technology OMEGA System [54], with the consent of: Jonathan Trent

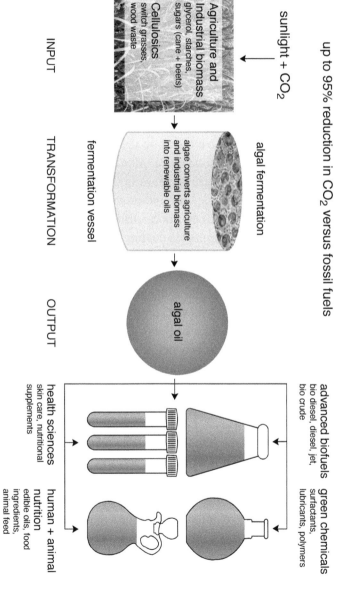

FIGURE 20: Schematic diagram drawing the production process of different industrial products by Solazyme technology [57]

9.6.6 PROCESSES OF SUNLESS PHOTOSYNTHESIS

Term of sunless photosynthesis may be used to define the process carried out by heterotrophic algae without the solar energy, but using the previously synthesized organic compounds as a source of carbon and energy.

Heterotrophic algae growth is dependent on many factors. A significant role in the stimulation or inhibition of heterotrophic microorganism's growth will play the access of sources organic compounds. Depending on the concentration as well as the amounts of organic compounds in a given nourishment, the algae biomass growth runs with different efficiencies. It was also found the positive effect on the growth of microorganisms in environment with poor carbon source mixture. Another factor that has an impact on the growth of algae biomass is the concentration of nutrients such as nitrogen. For example, the studies show the most effective source of nitrogen for the growing of microorganisms on the glucose nourishment is urea [56]. One of the companies dealing with the issue of sunless photosynthesis of algae on an industrial scale is a private company Solazyme Inc., which was found out in 2003. Method for waste biomass change-over to liquid fuels, developed in the company, uses various strains of heterotrophic genetically changed algae to produce different types of fuels. Figure 20 shows the main production pathway in Solazyme technology:

Alpha Algae, used by Solazyme, are heterotrophic organisms, which means they are developing without solar energy. Identified strains of algae, as it was shown in Figure 20, are transferred into the stainless steel containers. There, in darkness algae are fed by nourishment in the form of various types of raw material. This is mostly lignocellulosic biomass, so all kinds of grass, wood chips, agricultural sludge, as well as other waste substances. Depending on the type of nourishment, and the breeding strain, the algae produce different products. Some of them produces lipids, others provide a rich mixture of hydrocarbons, similar to those included in the light oil [58].

The biggest achievement of California's company happened on 13 March 2012. The ship LM 2500 fleet of U.S. Navy frigate, sailed successfully from the port in Everett (WA), to San Diego (CA), powered by fuel: Soladiesel HRD-76 ®. The ship was driven using 25,000 liters of 50/50 mixture of Soladiesel ® and pertoleum F-76. That was the first ever demonstration of alternative fuel mixture in the ship of naval fleet [59].

9.7 BIOHYDROGEN AND LIQUID ORGANIC HYDRIDES (LOH) APPLICATION

In the context discussed above solar fuels effective production of hydrogen could be the chance for application of LOH as chemical hydrogen storage medium.. The use of liquid organic hydrides in hydrogen storage provides high gravimetric and volumetric hydrogen density, low potential risk, and low capital investment because it is largely compatible with the current transport infrastructure, see Fig.21. Despite its technical, economical, and environmental advantages, the idea of hydrogen storage in liquid organic carriers has not been commercially proved yet. This is because of technical limitations related to the amount of energy required to extract the hydrogen from liquid organic hydride and dehydrogenation catalyst is not stable enough. Renewable hydrogen as well aromatic compounds can be obtained by direct catalytic conversion of methane or biomethane in dehydrocondensation (DHC) reaction. DHC processwas intensively was studied for the last decade by several groups of researchers, mostly from Japan [60-63], China [64], USA [55-67], and Hungary [68]. Following endothermic reaction represents methane direct conversion to benzene and hydrogen without participation of oxygen:

$$6 \, CH_4 \leftrightarrows C_6H_6 + 9H_2, \quad DH298 = 88,4 \, kJ/mol$$

The most effective catalyst for this process is Mo/H-ZSM-5 and Re/H-ZSM5. Oxygen-free conditions used for this reaction result in high benzene selectivity (up to 80%). Nowdays, methane is mostly used for heating purposes, as a transport fuel (CNG) and for chemical synthesis. DHC is promising process from point of view petrochemical feed stocks synthesis, hydrogen production for fuel cells and possible conversion of the waste and difficult accessible resources of natural gas. It can be clathrate (methane hydrate), coal bed methane, post fermentation biogas, land fill and recently shale gas able to be converted into liquid fuels, chemicals easily transportable liquid products.

9.8 ENVIRONMENTAL CONDITIONS FOR XTL PROCESSES IMPLEMENTATION

"XtL" processes include an indirect path of energy generation from different types of biomass by converting it to liquid energy carriers [70]. XtL processes means "x to liquid fuels," that is the conversion of various types of raw materials as feedstock for liquid biofuels. The previously presented innovative technologies of waste carbon dioxide conversion and other forms of organic matter into liquid energy carriers are part of a convention of XtL processes.

1. The environmental aspects of XtL technologies are related with:
2. Emission to the atmosphere the additional amounts of GHG.
3. Generating, recycling and removal of waste.
4. The use of natural resources, raw materials and energy.
5. Sludge production.
6. Space planning (development of the new areas on the Earth).
7. Impact on biodiversity.
8. The destruction of animal species and their habitats.

These methods are directed to the reuse, recycling of existing CO_2 for its processing into useful forms of energy. Algae used for production of liquid fuels are characterized by an effective reduction of significant amount of wastewater, through their purification from toxic substances such as heavy metals. The result of this microorganisms activity are the extra amounts of clean drinking water. It also shows that they have no negative impacts for ecosystems biodiversity, there are also not the reason of the disappearance of certain group of animals or plants.

As it is clear from the review of literature, cultivation of algae for fuels production, seems to be a forward-looking technology for producing liquid energy carriers. The algae have many important advantages from the environmental point of view:

• they are able to sequester waste carbon dioxide, which is used for production of oil and sugars, that are a substrate for the biofuels production,

- they have the ability to bioremediation of contaminated environments,
- they create a "biofertilizers" by absorbing elements such as nitrogen and phosphorus,
- they do not compete with the food market,
- they have a greater oil production capacity than terrestrial energy plants (more than 10 times greater than cultivated oilseed crops),
- they are a group of organisms, with possibility of genetic changes for improvement of their production ability.

Concluding, it is clear from the above considerations, the discussed projects may be an excellent alternative to conventional resources of crude oil, gas and coal. Finally by reducing the concentration of carbon dioxide in the atmosphere, without any negative impact on the environment, above projects have large potential for limitation so-called Global Warming.

9.9 SWOT ANALYSIS OF SELECTED TECHNOLOGIES

9.9.1 TECHNOLOGY OF "ARTIFICIAL LEAF"

The strength of "artificial leaf" is ability to carry out the hydrolysis process of water molecule into oxygen and hydrogen, which is the cleanest known fuel. It cannot be possibly limited in future only for use pure water but also polluted one. In this way, perhaps in the future could be used even salty water. Another strong point of this method is the efficiency of the artificial leaf, which now comes to (20... 30)%. If Artificial Leaf would be able to bind waste carbon dioxide from the exhausted gases then it can be an effective method of producing so-called solarfuels.

9.9.2 METHANOL SYNTHESIS IN CARBON DIOXIDE PHOTOREDUCTION PROCESS

The opportunity for the development of EtL processes is a partnership with institutions leading the projects of ethanol production of all kinds of waste material.

9.9.3 THE PROCESS OF PHOTOSYNTHESIS WITHOUT SUNLIGHT

The advantage of sunless photosynthesis is the genetic modification of the strains cultivated algae. Genetic engineering allows increasing of algae productivity, and efficient development of microbial species without the solar energy. This in turn results in the elimination of risks in the form of competition from autotrophic algae.

9.9.4 OMEGA SYSTEM – MARINE MEMBRANE ALGAE CULTIVATION FARMS

Strong side of OMEGA system is the raw material for fed of algae. This allows to use the opportunity that comes from the cultivation of algae for example, between San Diego and San Francisco, where is discharged from industrial plants about 7 billion gallons per day of sewage into the sea. This would be a very effective method of disposing of waste, and even that would be a form of recycling, as a result of microbial activity would arise alternative fuels.

9.10 CONCLUSIONS

Cultivation of oil plants for fuels production creates strong competition with crops for consumption. It's hard to talk about the promotion of bio-fuels, which production is based on energy crops and in many cases is a major cause of the food crops shortage. In many regions of the World are starving societies of so-called third world countries. Therefore as the most desirable features of technology is to use waste CO_2 to produce liquid fuels related with:

- reality;
- affordability;
- efficiency;
- the environmental aspect.

In addition, all products and waste that may be generated during the manufacturing process for liquid energy carriers, should use the strategy 3R—Recycling Reduce Reuse in order to reduce energy costs and protect the environment. It is believed that the process of hydrolysis of water by solar energy, in the artificial leaf technology, occurs and is energy efficient.

Harry B. Gray, winner of many awards, including in chemistry, in an article on "Solar Fuel" says:

We need to stop burning hydrocarbon as soon as it is possible, because they are wonderful resources that we desperately need for the production of dyes, pharmaceuticals, T-shirts, chairs and cars —is irresponsible to burn them![71].

Waste Carbon dioxide is appropriate material. It is a resource that exists in plenty and causing a negative impact on the environment by including contributing to the greenhouse effect. Artificial photosynthesis is in the research phase and need more decades to make it economically profitable. Today we can see many successful trials to transform CO_2 into various liquid fuel.

The presented analysis shows, that the most important goals for the energy sector should be utilization of carbon dioxide emissions by mentioned CCU idea. This target can be realized by using modern technology of CO_2 reuse to produce liquid energy carriers.

REFERENCES

1. What the Future Holds in Store, in: World Climate Report, The Web's Longest-Running Climate Change Blog, 30 January, 2008, available from: http://www.worldclimatereport.com/index.php/2008/01/30/what-the-future-holds-in-store/ (accessed 20 August 2012)
2. CO2 emissions from fuel combustion IEA Statistics, available from: http://www.iea.org/co2highlights/CO2highlights.pdf (accessed 12 October 2011)
3. W jaki sposób świeci Słońce?, available from: http://urania.pta.edu.pl/bahcall1.html dn.12.12.2011 (accessed 25 October 2011) ("How does the Sun shines?")
4. Van Loon G., Duffy S. J., Gazy cieplarniane związane ze stosowaniem paliw opartych na węglu; Bilans energetyczny; Chemia klimatu globalnego, in Chemia środowiska, Pub. PWN SA, Warsaw, 2007. (Greenhouse gases associated with the use of carbon based fuels; The energy balance; Chemistry of the global climate, in Environmental chemistry)

5. Solar energy distribution graph showing that infrared radiation makes up a large portion of the solar spectrum, available from: http://continuingeducation.construction. com/article.php?L=68&C=488&P=6 (accessed 24 October 2012)

6. Solar Spectrum, from Wikipedia, the free encyclopedia, available from: http:// en.wikipedia.org/wiki/File:Solar_Spectrum.png (accessed 24 April 2012)

7. Płaska J., Sałek M., Surma T., Wykład pierwszy. Charakterystyka odnawialnych źródeł energii, „Energetyka" – March 2005. („The first lecture. Characteristics of renewable sources of energy")

8. Strebeyko P., Fotosynteza, Pub. PWN, Warsaw, 1964. ("Photosynthesis")

9. Staroń K., (ed.) Przetwarzanie energii i materii podczas fotosyntezy, in Biologia, Pub. WSiP, Warsaw, 2003, ISBN 83-02-08628-2) pp. 106-120. ("The processing of energy and matter during photosynthesis in Biology")

10. Luminos output of grow lights is not the right parameter, available from: http:// www.myledlightingguide.com/Article.aspx?ArticleID=39 (accessed 24 April 2012)

11. Burda K., Dlaczego warto zajmować się fotosyntezą?, FOTON 93, Summer 2006, available from: http://www.if.uj.edu.pl/Foton/93/pdf/04%20fotosynteza.pdf (accessed 20 September 2012) („Why is photosynthesis?")

12. Directive 2009/29/EC, available from: http://eur-lex.europa.eu/LexUriServ/Lex-UriServ.do?uri=OJ:L:2009:140:0016:0062:pl:PDF (accessed 12 October 2011)

13. Beret-Kowalska G., Kacprowska J., et al., Energia ze źródeł odnawialnych w 2007 roku, Information and Statistical Papers, Central Statistical Office, Department of Industry, Ministry of Economy, Department of Energy, Warsaw, 2008, pp. 5-14. (ISSN: 1898-4347) ("Energy from renewable sources in 2007")

14. Grzybek A., Analiza możliwości wykorzystania surowców rolnych na potrzeby produkcji biopaliw, „Problemy inżynierii Rolniczej", No. 1/2007. („Analysis of agricultural raw materials for the production of biofuels")

15. Grzybek A., Zapotrzebowanie na biomasę i strategie energetycznego jej wykorzystania, „Studies and reports IUNG – PIB", Issue II, 2008, Warsaw.) („The demand for biomass energy and its utilization strategies")

16. Lipski R., Orliński S., Tokarski M., Energetyczne wykorzystanie biomasy na przykładzie kotłowni opalanej słomą we Fromborku, MOTROL, 2006, 8A, pp. 202-209. ("Energy use of biomass on an example of straw-fired boiler in Frombork")

17. Roszkowski A., Efektywność energetyczna różnych sposobów produkcji i wykorzystania biomasy, the Institute for Building, Mechanization and Electrification of Agriculture, Warsaw, Studies and Reports IUN, issue 11/2008, available from: http://217.113.158.23/wydawnictwa/Pliki/pdfPIB/zesz11.pdf#page=100 (accessed 12 April 2012) („The effectiveness of different methods of biomass production and use")

18. Biopaliwa – ratunek dla klimatu czy zagrożenie dla przyrody, Warsaw, 5 April 2011, OTOP, available from: http://otop.org.pl/uploads/media/informacja_prasowa_biopaliwa.pdf (accessed 23 March 2012) („Biofuels - the rescue of the climate or the threat to wildlife?")

19. Wójcicki Z., Energia odnawialna, biopaliwa i ekologia, IBMiER in Warsaw, „Problemy Inżynierii Rolniczej" No. 2/2007. ("Renewable energy, biofuels and ecology")

20. Niedziółka I., Zuchniarz A., Analiza energetyczna wybranych rodzajów biomasy pochodzenia roślinnego, Department of Agricultural Machinery, Agricultural Uni-

versity of Lublin, „MOTROL", 2006, 8A, pp.233, available form: http://www.pan-ol.lublin.pl/wydawnictwa/Motrol8a/Niedziolka.pdf (accessed 23 April 2012) ("Energy analysis of selected types of plant biomass")

21. Kijeńska M., Kijeński J., Biopaliwa czy wyczerpaliśmy już wszystkie możliwości?, „CHEMIK", 2011, available from: http://chemia.wnp.pl/biopaliwa-czy-wyczerpalismy-juz-wszystkie-mozliwosci,7511_2_0_2.html (accessed 22 October 2011) („Biofuels – did we exhauste all the possibilities?")

22. Taylor K., Ivanovsky W., Spirits mixture of diesel, "Chemical Industry" No. 11-12, Lvov, Poland, 1926

23. http://ec.europa.eu/clima/policies/roadmap/index_en.htm (accessed 09 October 2012)

24. Technology Roadmap Biofuels for Transport © OECD/IEA, 2011, International Energy Agency 9 rue de la Fédération 75739 Paris Cedex 15, France

25. Bosmans A., Vanderreydt I., Geysen D., Helsen L., The crucial role of Waste-to-Energy technologies in enhanced landfill mining: a technology review, Journal of Cleaner Production. In press. Available from: http://www.sciencedirect.com/science/article/pii/S0959652612002557(accessed 19 September 2012)

26. Naber J. E., Goudriaan F., (BIOFUEL BV), Successfully using biomass to Harness renewable energy in an efficient and cost-effective way, available from: https://docs. google.com/viewer?a=v&q=cache:_SpnMHqVSawJ:www.cpi.umist.ac.uk/eminent/Confidential/meeting/RigaMeeting/Riga%2520Workshop/PresenatieHTUBiofuel.ppt+&hl=en&gl=pl&pid=bl&srcid=ADGEESgMHqRXurQxLkzf8s6voe2T2A smrxLMOGwHrqmqRYlcUMMekL57T8pTe-LtMDgxMN6dLSYz30U37W5gpDo-AKfAISwapjspGTUrVB9XwICS5SkcnnJ0uka5wrwNfz6AD27qedYLa&sig=AHI EtbTXFACuXwfDtR9YJho3R2xu1V8VTw (accessed 9 September 2012)

27. Li G., Kong L., Wang H., Huang J., Xu J., Application of hydrothermal reaction in resource recovery of organic wastes, State Key Lab of Pollution Control and Resource Reuse, School of Environmental Science and Engineering, Tongji University, No. 1239 Siping Road, Shanghai 200092, China

28. Malinowski A., Perspektywiczne technologie biopaliw drugiej generacji na rynku europejskim, conference: „Rynek biopaliw w unijnym pakiecie dot. energii i klimatu", Warsaw, 27 September 2010 („Promising second-generation biofuels technologies in the European market")

29. Official website of Alphakat Diesel: http://www.alphakatdiesel.pl/technologiakdv. html (accessed 19 April 2012)

30. Technology modules of KDV, available from: http://www.kdv-projektmanagement. com/technology (accessed 19 April 2012)

31. Official website of AlphakatPoland: http://alphakatpolska.pl/char_pro.html (accessed 19 April 2012)

32. Official website of JCAP: http://solarfuelshub.org/hub-overview

33. PL Patent 208030 B1, „Katalizator do syntezy metanolu i jego pochodnych" by Nazimek D., Czech B., Jabłoński B., Zaniuk W., Wasińska Z. ("The catalyst for the synthesis of methanol and its derivatives").

34. Nazimek D., (2009) Sztuczna Fotosynteza CO2 - pierwszy etap produkcji biopaliw, available from: http://www.gmina.zgierz.pl/cms/files/File/KONFERENCJA%20

Rogozno%20-%20wczoraj,%20dzis,%20jutro/SFCO2-2009-Krakow.pdf ("Artificial Photosynthesis CO2 - the first stage in the production of biofuels").

35. Report: Polish Power Plant and University to Cooperate on CO2 to Methanol Trial, available from: http://www.greencarcongress.com/2009/07/nazimek-20090707. html (accessed 21 April 2012)

36. Nazimek D., Niećko J., (2010) Coupling Ethanol with Synthetic Fuel, Polish J. of Environ. Stud. Vol. 19, No. 3 (2010), 507-514, available from: http://www.pjoes. com/pdf/19.3/507-514.pdf (accessed 20 September 2012)

37. MIT's Daniel Nocera Announces Artificial Leaf With Goal To Make Every Home a Power Station, Signs with Tata, available from: http://www.freeenergytimes. com/2011/03/28/mit-chemist-daniel-nocera-announces-artificial-leaf-goal-to-make-every-home-a-power-station/ (accessed 10 May 2012)

38. Van Noorden R., Secrets of artificial leaf revealed, Nature, doi:10.1038/ news.2011.564, published online 29 September 2011, available from: http://www. nature.com/news/2011/110929/full/news.2011.564.html (accessed 10 May 2012)

39. Chandler D. L., Taking a Leaf from Nature's Book, Technology Review published by MIT, January/February 2012, available from: http://www.technologyreview.com/ printer_friendly_article.aspx?id=39259 (accessed 24 April 2012)

40. Regalado A., „Reinventing the Leaf. The ultimate fuel may come not from corn or algae but directly from sun itself.", „Scientific American", October 2010, pp.68-71.

41. Prototype reactor uses the sun to generate storable hydrogen fuel , available from: http://www.mobilemag.com/2010/12/28/prototype-reactor-uses-the-sun-to-generate-storable-hydrogen-fuel/ (accessed 24 April 2012)

42. Chueh W. C., Falter C., Abbott M., Scipio D., Furler P., Haile S. M., Steinfeld A., High-Flux Solar-Driven Thermochemical Dissociation of CO2 and H2O Using Nonstoichiometric Ceria, „Science", 24.12.2010.: Vol. 330 No. 6012 pp.1797-1801.

43. Scott S. A., Biodiesel from alga: challenges and prospects, „ScienceDirect", Current Opinion in Biotechnology 2010,21:277-286.

44. Fishman D., Majumdar R., Morello J., Pate R., National Algal Biofuels Technology Roadmap, U.S. Department of Energy, Office of Energy Efficiency and Renewable Energy, Biomass Program, U.S. DOE 2010.

45. Burczyk B., Biomasa. Surowiec do syntez chemicznych i produkcji paliw, OWPW, Wrocław 2011, pp.180,244. („Biomass. Raw material for the chemical synthesis and the production of fuels")

46. Frąc M., et al, Algi – energia jutra (biomasa, biodiesel), Acta Agrophysica 2009, 13(3), 627-638, Instytut Agrofizyki im. Bohdana Dobrzańskiego PAN, Lublin, pp.634. („Algae - tomorrow's energy (biomass, biodiesel")

47. Zakrzewski T., Biomasa mikroalg-obiecujące paliwo przyszłości, „Czysta Energia", 2/2011(114). ("Biomass microalgae-promising fuel of the future", "Clean Energy")

48. Kozieł W., Włodarczyk T., Glony-produkcja biomasy, Acta Agrophysica 2011, 17(1), 105-116, Instytut Agrofizyki im. Bohdana Dobrzańskiego PAN, Lublin, pp. 109. („Algae-biomass production")

49. Cieśliński H., Biodiesel z alg – paliwo przyszłości?, available from: http://www. silawiedzy.pl/images/stories/artykuly_pdf/837.pdf (accessed 12 April 2012) ("Biodiesel from algae - fuel of the future?")

50. Sapphire Energy technology, available from: http://www.sapphireenergy.com (accessed 11 April 2012)

51. Malinowski A., Biopaliwa dla Lotnictwa (Aviation Biofuels), Czysta Energia (Clean Energy), Nr: 121, 9/2011; http://e-czytelnia.abrys.pl/index.php?mod=tekst&id=13518 (accessed 08 October 2012)

52. Algenol's carbon platform modular technology, available from: http://www.algenolbiofuels.com/direct-to-ethanol/direct-to-ethanol (accessed 23 April 2012)

53. OMEGA, in AlgaeSystems, available from: http://algaesystems.com/technology/omega/ (accessed 24 April 2012)

54. Schwartz D., AIM Interview: NASA's OMEGA Scientist, Dr. Jonathan Trent, August 21, 2011in Algae Industry Magazine website, available from: http://www.algaeindustrymagazine.com/nasas-omega-scientist-dr-jonathan-trent/ (accessed 24 April 2012)

55. Trent J., Wiley P., Tozzi S., McKuin B., Reinsch S., Biofuels, Volume 3, Issue 5, The future of biofuels: is it in the bag?, pp. 521-524, 01 September, 2012

56. Sikora Z., Heterotroficzny wzrost glonów, „Wiadomości Botaniczne", Volume XIII – Issue 2, 1969. ("Heterotrophic growth of algae")

57. Gunther M., Solazyme's amazing algae, available from: http://www.marcgunther.com/2010/03/18/solazymes-amazing-algae/ (accessed 11 April 2012)

58. Fuel from algae, available from: www.technologyrewiev (accessed 24 August 2011)

59. U.S. Navy Successfully Tests SoladieselHRD-76® During Operational Transit Voyage in USS Ford Frigate, Solazyme Inc., March 13, 2012, available from: http://investors.solazyme.com/releasedetail.cfm?ReleaseID=656801 (accessed 13 March 2012)

60. Ohnishi R., Liu S., Dong Q., Wang L., Ichikawa M., Catalytic Dehydrocondensation of Methane with CO and CO2 toward Benzene and Naphthalene on Mo/HZSM-5 and Fe/Co-Modified Mo/HZSM-5, J. Catal. 182, 92 (1999).

61. Shu Y., Ma H., Ohnishi R. and Ichikawa M., Highly stable performance of catalytic methane dehydrocondensation towards benzene on Mo/HZSM-5 by a periodic switching treatment with H2 and CO2, Chem. Commun., 86, (2003).

62. Malinowski A., Ohnishi R., Ichikawa M., CVD Synthesis in Static Mode of Mo/H-ZSM5 Catalyst for the Methane Dehydroaromatization Reaction to Benzene, Catalysis Letters Vol.96, No 3-4, pp 141-146, (2004)

63. Patent JP2002336704 Japan Steel Works LTD, Aromatizing reaction catalyst of methane and method for preparing the same, 26 Nov. 2002, Ichikawa M., Ohnishi R., Malinowski A.,. Ikeda K., Yagi K.; The European Patent Office: http://ep.espacenet.com

64. Shu Y., Xu Y., Wong S., Wang L., Guo X., Promotional Effect of Ru on the Dehydrogenation and Aromatization of Methane in the Absence of Oxygen over Mo/HZSM-5 Catalysts, J. Catal. 170 11 (1997).

65. Weckhuysen B.M., Wang D., Rosynek M.P., and Lunsford J.H., Conversion of Methane to Benzene over Transition Metal Ion ZSM-5 Zeolites: II. Catalyst Characterization by X-Ray Photoelectron Spectroscopy ,J. Catal. 175, 347 (1998).

66. Wang D., Lunsford J.H., Rosynek M.P., Characterization of a Mo/ZSM-5 Catalyst for the Conversion of Methane to Benzene, J. Catal. 169 347 (1997).

67. Solymosi F., Cserényi J., Szöke A., Bánsagi T., and Oszkó A. J., Aromatization of Methane over Supported and Unsupported Mo-Based Catalysts ,J. Catal. 165, 150 (1997).
68. Liu Z., Nutt M.A. and Iglesia E., The Effects of CO2, CO and H2 Co-Reactants on Methane Reactions Catalyzed by Mo/H-ZSM-5, Catal. Lett., 2002, 81, 271-279.
69. PACIFICHEM Conference 2005, #872
70. Biernat K., Biokomponenty i biopaliwa – możliwości rozwoju i zastosowania, „Forum Czystej Energii" X edycja, 23-25 listopada 2010 r., Poznań, „Technology Roadmap. Biofuels for Transport" Pub. OECE/IEA, Paris 2011. („Biocomponents and biofuels - opportunities for development and use")
71. Gray, H. B., Solar Fuel, „Engineering and Science" 60 (3). pp. 28-33, 1997, ISSN 0013-7812, available from: http://calteches.library.caltech.edu/3924/1/Fuel.pdf (accessed 20 September 2012)

There are several tables that are not available in this version of the article. To view this additional information, please use the citation on the first page of this chapter.

CHAPTER 10

Well-to-Wheels Energy Use and Greenhouse Gas Emissions of Ethanol from Corn, Sugarcane, and Cellulosic Biomass for US Use

MICHAEL WANG, JEONGWOO HAN, JENNIFER B. DUNN, HAO CAI, AND AMGAD ELGOWAINY

10.1 INTRODUCTION

Globally, biofuels are being promoted for reducing greenhouse gas (GHG) emissions, enhancing the domestic energy security of individual countries and promoting rural economic development. In a carbon-constrained world, liquid transportation fuels from renewable carbon sources can play an important role in reducing GHG emissions from the transportation sector (IEA 2012). At present, the two major biofuels produced worldwide are (1) ethanol from fermentation of sugars primarily in corn starch and sugarcane and (2) biodiesel from transesterification of vegetable oils, with ethanol accounting for the majority of current biofuel production. Figure 1 shows the growth of annual ethanol production between 1981 and 2011 in the US and Brazil, the two dominant ethanol-producing countries.

Well-to-Wheels Energy Use and Greenhouse Gas Emissions of Ethanol from Corn, Sugarcane, and Cellulosic Biomass for US Use. © *Wang M, Han J, Dunn JB, Cai H, and Elgowainy A.* Environmental Research Letters *7,4 (2012), doi:10.1088/1748-9326/7/4/045905. Licensed under a Creative Commons Attribution 3.0 Generic License, http://creativecommons.org/licenses/by/3.0/.*

Figure 1 from Michael Wang et al 2012 Environ. Res. Lett. 7 045905

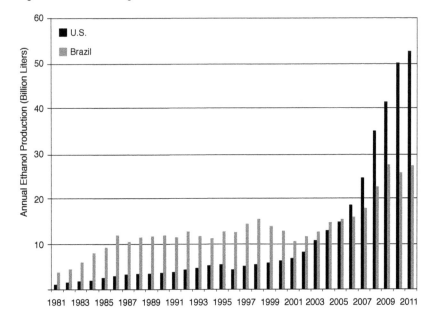

FIGURE 1: Annual ethanol production in the US and Brazil (based on data from the Renewable Fuels Association (RFA 2012) and Brazilian Sugarcane Association (UNICA 2012)).

The production of corn ethanol in the US has increased to more than 52 billion liters since the beginning of the US ethanol program in 1980. The increase after 2007, the year the Energy Independence and Security Act (EISA) came into effect, is remarkable. Growth in the production of Brazilian sugarcane ethanol began in the 1970s when the Brazilian government began to promote its production. The most recent growth in sugarcane ethanol production, since 2001, has mainly resulted from the popularity of ethanol flexible-fuel vehicles and from the advantageous price of ethanol over gasoline in Brazil.

Over the long term, the greatest potential for bioethanol production lies in the use of cellulosic feedstocks, which include crop residues (e.g., corn stover, wheat straw, rice straw and sugarcane straw), dedicated energy crops (e.g., switchgrass, miscanthus, mixed prairie grasses and short-rotation trees) and forest residues. The resource potential of these cellulosic feedstocks can support a huge amount of biofuel production. For example, in the US, nearly one billion tonnes of these resources are potentially available each year to produce more than 340 billion liters of ethanol per year (DOE 2011). This volume is significant, even when compared to the annual US consumption of gasoline, at 760 billion ethanol-equivalent liters (EIA 2012).

The GHG emission reduction potential of bioethanol, especially cellulosic ethanol, is recognized in policies that address reducing the transportation sector's GHG emissions (i.e., California's low-carbon fuel standard (LCFS; CARB 2009), the US renewable fuels standard (RFS; EPA 2010) and the European Union's renewable energy directive (RED; Neeft et al 2012)). Nonetheless, the life-cycle GHG emissions of bioethanol, especially those of corn-based ethanol, have been subject to debate (Farrell et al 2006, Fargione et al 2008, Searchinger et al 2008, Liska et al 2009, Wang et al 2011a, Khatiwada et al 2012). With regard to corn ethanol, some authors concluded that its life-cycle GHG emissions are greater than those from gasoline (Searchinger et al 2008, Hill et al 2009). Others concluded that corn ethanol offers reductions in life-cycle GHG emissions when compared with gasoline (Liska et al 2009, Wang et al 2011a). On the other hand, most analyses of cellulosic ethanol reported significant reductions in life-cycle GHG emissions when compared with those from baseline gasoline. Reductions of 63% to 118% have been reported (Borrion et al 2012, MacLean and Spatari 2009, Monti et al 2012, Mu et al 2010, Scown et al 2012, Wang et al 2011a, Whitaker et al 2010). Most of these studies included a credit for the displacement of grid electricity with electricity co-produced at cellulosic ethanol plants from the combustion of lignin. Some, however, excluded co-products (e.g., MacLean and Spatari 2009). Uniquely, Scown et al (2012) considered land use change (LUC) GHG emissions (for miscanthus ethanol) and estimated total net GHG sequestration of up to 26 g of CO_2 equivalent (CO_2e)/MJ of ethanol. In the case of sugarcane ethanol, Seabra et al (2011) and Macedo et al (2008) reported life-cycle GHG emissions that were between 77% and 82% less than those of baseline gasoline. Wang et al (2008) estimated this reduction to be 78%.

A detailed assessment of the completed studies requires that they be harmonized with regard to the system boundary, co-product allocation methodology, and other choices and assumptions that were made. Other researchers (e.g., Chum et al 2011) have undertaken this task to some extent. Here we instead use a consistent modeling platform to examine the GHG impacts from using corn ethanol, sugarcane ethanol and cellulosic ethanol. The GREET (Greenhouse gases, Regulated Emissions, and Energy use in Transportation) model that we developed at Argonne National Laboratory has been used by us and many other researchers to examine GHG emissions from vehicle technologies and transportation fuels on a consistent basis (Argonne National Laboratory 2012). The GREET model covers bioethanol production pathways extensively; we have updated key parameters in these pathways based on recent research. This article presents key GREET parametric assumptions and life-cycle energy and GHG results for bioethanol pathways contained in the GREET version released in July 2012. Moreover, we quantitatively address the impacts of critical factors that affect GHG emissions from bioethanol.

10.2 SCOPE, METHODOLOGY, AND KEY ASSUMPTIONS

We include bioethanol production from five feedstocks: corn grown in the US, sugarcane grown in Brazil, and corn stover, switchgrass and miscanthus, all grown in the US. Even though the wide spread drought in the US midwest in the summer of 2012 may dampen corn ethanol production in 2012, corn ethanol production will continue to grow, possibly exceeding the goal of 57 billion liters per year in the 2007 EISA. Likewise, Brazil's sugarcane ethanol production will continue to grow. In the US midwest corn belt, up to 363 million tonnes of corn stover can be sustainably harvested in a year (DOE 2011). Large-scale field trials have been in place to collect and transport corn stover (Edgerton et al 2010). Switchgrass is a native North American grass. Field trials of growing switchgrass as an energy crop have been in place since the 1980s. Miscanthus, on the other hand, has a high potential yield per acre. In the past several years, significant efforts have been made in the US to develop better varieties of miscanthus with higher yields (Somerville et al 2010).

Figure 2 from Michael Wang et al 2012 Environ. Res. Lett. 7045905

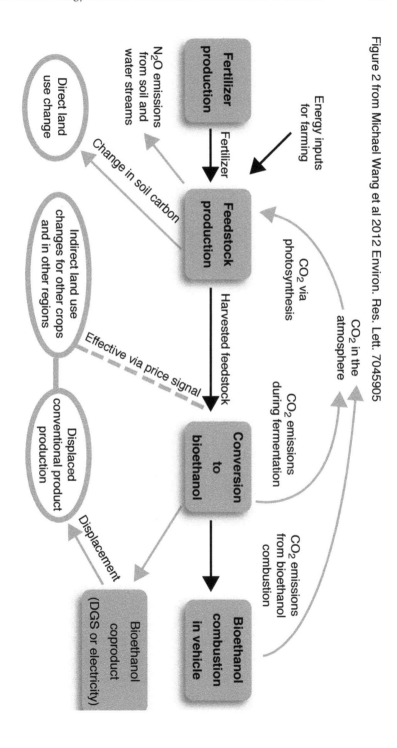

FIGURE 2: System boundary of well-to-wheels analysis of bioethanol pathways.

We conducted the well-to-wheels (WTW, or, more precisely for bio-ethanol, field-to-wheels) analyses of the five bioethanol pathways with the GREET model (Argonne National Laboratory 2012, Han et al 2011, Dunn et al 2011, Wang et al 2012). In particular, we used the most recent GREET version (GREET1_2012) for this analysis to conduct simulations for the year 2015. Figure 2 presents the system boundary for the five bio-ethanol pathways in our analysis. Parametric details of the five pathways are presented below. For comparison, we included petroleum gasoline in our analysis.

The GREET model is designed with a stochastic modeling tool to address the uncertainties of key parameters and their effects on WTW results. For this article, we used that feature to conduct simulations with probability distribution functions for key parameters in the WTW pathways. In addition, we conducted parametric sensitivity analyses to test the influence of key parameters on GHG emissions for each of the five pathways.

10.2.1 CORN-TO-ETHANOL IN THE US

For the corn-to-ethanol pathway, corn farming and ethanol production are the two major direct GHG sources (Wang et al 2011a). From farm-ing, N_2O emissions from the nitrification and denitrification of nitro-gen fertilizer in cornfields, fertilizer production and fossil fuel use for farming are significant GHG emission sources. GHG emissions during ethanol production result from the use of fossil fuels, primarily natural gas (NG), in corn ethanol plants. GREET takes into account GHG emis-sions from NG production and distribution (such as methane leakage during these activities (see Burnham et al 2012)) as well as those from NG combustion. The treatment of distillers' grains and solubles (DGS), a valuable co-product from corn ethanol plants, in the life-cycle analysis (LCA) of corn ethanol is important because it can affect results regard-ing corn ethanol's GHG emissions (Wang et al 2011b). Table 1 presents key parametric assumptions in GREET for corn-based ethanol. In this and subsequent tables, P10 and P90 represent the 10th and 90th percen-tiles, respectively, of these parameters.

TABLE 1: Parametric assumptions about the production of ethanol from corn in the US.

Parameter: unit	Mean	P10	P90	Distribution function type
Corn farming: per tonne of corn (except as noted)				
Direct energy use for corn farming: MJ	379	311	476	Weibull[a]
N fertilizer application: kg	15.5	11.9	19.3	Normal[a]
P fertilizer application: kg	5.54	2.86	8.61	Lognormal[a]
K fertilizer application: kg	6.44	1.56	12.5	Weibull[a]
Limestone application: kg	43.0	38.7	47.3	Normal[a]
N_2O conversion rate of N fertilizer: %	1.525	0.413	2.956	Weibull[b]
NG use per tonne of ammonia produced: GJ	30.7	28.1	33.1	Triangular[c]
Corn ethanol production				
Ethanol yield: l/tonne of corn	425	412	439	Triangular[a]
Ethanol plant energy use: MJ/l of ethanol	7.49	6.10	8.87	Normal[a]
DGS yield: kg (dry matter basis)/l of ethanol	0.676	0.609	0.743	Triangular[a]
Enzyme use: kg/tonne of corn	1.04	0.936	1.15	Normal[d]
Yeast use: kg/tonne of corn	0.358	0.323	0.397	Normal[d]

[a]The type and shape of distribution functions were developed in Brinkman et al (2005). The means of the distributions were scaled later to the values in Wang et al (2007, 2011a). [b]Based on our new assessment of the literature, see supporting information (available at stacks.iop.org/ERL/7/045905/mmedia) for details. [c]From Brinkman et al (2005). [d]Selected among 11 distribution function types, with maximization of the goodness-of-fit method to the data compiled in Dunn et al (2012a).

10.2.2 PRODUCTION OF ETHANOL FROM SUGARCANE IN BRAZIL FOR USE IN THE US

Brazilian sugarcane mills produce both ethanol and sugar, with the split between them readily adjusted to respond to market prices. Bagasse, the residue after sugarcane juice is squeezed from sugarcane, is combusted in sugar mills to produce steam (for internal use) and electricity (for internal use and for export to the electric grid). Sugarcane farming is associated with significant GHG emissions from both upstream operations such as fertilizer production and from the field itself. For example, the nitrogen (N) in sugarcane residues (i.e., straw) on the field as well as the N in fertilizer emit N_2O. The sugar mill by-products vinasse and filter cake applied as soil amendments also emit N_2O as a portion of the N in them degrades (Braga do Carmo et al 2012). Open field burning, primarily with manual harvesting of sugarcane (which is being phased out), and transportation logistics (truck transportation of sugarcane from fields to mills and of ethanol from mills to Brazilian ports; ocean tanker transportation of ethanol from southern Brazilian ports to US ports; and US ethanol transportation) are also key GHG emission sources in the sugarcane ethanol life cycle. Table 2 lists key parametric assumptions for the sugarcane-to-ethanol pathway. We did not have data on enzyme and yeast use for sugarcane ethanol production, so their impacts are not considered in this analysis. Given that enzymes and yeast have a minor impact on corn ethanol WTW results (Dunn et al 2012a), we expect that their effect on sugarcane WTW results are small as well.

10.2.3 CORN STOVER-, SWITCHGRASS- AND MISCANTHUS-TO-ETHANOL

The yield of corn stover in cornfields could match corn grain yield on a dry matter basis. For example, for a corn grain yield of 10 tonnes (with 15% moisture content) per hectare, the corn stover yield could be 8.5 tonnes (bone dry) per hectare. Studies concluded that one-third to one-half of corn stover in cornfields can be sustainably removed without causing erosion or deteriorating soil quality (Sheehan et al 2008, DOE 2011). When stover is removed, N, P and K nutrients are removed, too. We assumed in GREET simulations that the amount of nutrients lost with stover removal would be supplemented with synthetic fertilizers. We developed our replacement rates based on data for nutrients contained in harvested corn stover found in the literature (Han et al 2011).

Switchgrass can have an annual average yield of 11–13 tonnes ha^{-1}, with the potential of more than 29 tonnes ha^{-1} (Sokhansanj et al 2009). To maintain a reasonable yield, fertilizer is required for switchgrass growth. In arid climates, irrigation may be also required. In our analysis, we assumed that switchgrass would be grown in the midwest, south and southeast US without irrigation. Miscanthus can have yields above 29 tonnes ha^{-1} (with up to 40 tonnes) (Somerville et al 2010). Similar to switchgrass, fertilizer application may be required in order to maintain good yields.

TABLE 2: Parametric assumptions about the production of sugarcane ethanol in Brazil and its use in the US (per tonne of sugarcane, except as noted).

Parameter: unit	Mean	P10	P90	Distribution function type
Sugarcane farming				
Farming energy use for sugarcane: MJ	100	90.2	110	Normal[a]
N fertilizer use: g	800	720	880	Normal[a]
P fertilizer use: g	300	270	330	Normal[a]
K fertilizer use: g	1000	900	1100	Normal[a]
Limestone use: g	5200	4680	5720	Normal[a]
Yield of sugarcane straw: kg	140	126	154	Normal[a]
Filter cake application rate: kg (dry matter basis)	2.87	2.58	3.16	Normal[a]
Vinasse application rate: l	570	513	627	Normal[a]
Share of mechanical harvest: % of total harvest	80	NA[b]	NA[b]	Not selected
N$_2$O conversion rate of N fertilizer: %	1.22	1.05	1.39	Uniform[c]
Sugarcane ethanol production				
Ethanol yield: l	81.0	73.1	89.0	Normal[a]
Ethanol plant energy use: fossil kJ/l of ethanol	83.6	75.3	92.0	Normal[a]
Electricity yield: kWh	75	57.8	100	Exponential[a]
Sugarcane ethanol transportation				
Ethanol transportation inside of Brazil: km	690	NA[b]	NA[b]	Not selected
Ethanol transportation from Brazil to the US: km	11930	NA[b]	NA[b]	Not selected

[a]By maximization of goodness-of-fit to the data in Macedo et al (2004, 2008) and Seabra et al (2011). [b]NA=not available. [c]Data on N$_2$O emissions from sugarcane fields is very limited, so we assumed uniform distribution. See supporting information (available at stacks.iop.org/ERL/7/045905/mmedia) for details.

In cellulosic ethanol plants, cellulosic feedstocks go through pretreatment with enzymes that break cellulose and hemicellulose into simple sugars for fermentation. The lignin portion of cellulosic feedstocks can be used in a combined heat and power (CHP) generator in the plant. The CHP generator can provide process heat and power in addition to surplus electricity for export to the grid. Ethanol and electricity yields in cellulosic ethanol plants are affected by the composition of cellulosic feedstocks (although we did not find enough data to identify the differences in ethanol and electricity yield for our study). Lignin can also be used to produce bio-based products instead of combustion. In our analysis, we assume combustion of lignin for steam and power generation. Table 3 presents key assumptions for the three cellulosic ethanol pathways.

10.2.4 LAND USE CHANGE FROM BIOETHANOL PRODUCTION

Since 2009, we have been addressing potential LUC impacts of biofuel production from corn, corn stover, switchgrass and miscanthus with Purdue University and the University of Illinois (Taheripour et al 2011, Kwon et al 2012, Mueller et al 2012, Dunn et al 2012b). We developed estimates of LUC GHG emissions with a GREET module called the Carbon Calculator for Land Use Change from Biofuels Production (CCLUB) (Mueller et al 2012). In CCLUB, we combine LUC data generated by Purdue University from using its Global Trade Analysis Project (GTAP) model (Taheripour et al 2011) and domestic soil organic carbon (SOC) results from modeling with CENTURY, a soil organic matter model (Kwon et al 2012) that calculates net carbon emissions from soil. Above ground carbon data in CCLUB for forests comes from the carbon online estimator (COLE) developed by the USDA and the National Council for Air and Stream Improvement (Van Deusen and Heath 2010). International carbon emission factors for various land types are from the Woods Hole Research Center (reproduced in Tyner et al (2010)). We provide a full analysis of CCLUB results for these feedstocks elsewhere (Dunn et al 2012b) and summarize them briefly here.

When land is converted to the production of biofuel feedstock, direct impacts are changes in below ground and above ground carbon content, although the latter is of concern mostly for forests. These LUC-induced changes cause SOC content to either decrease or increase, depending on the identity of the crop. For example, if land is converted from cropland-pasture to corn, SOC will decrease, and carbon will be released to the atmosphere. However, conversion of this same type of land to miscanthus or switchgrass production likely sequesters carbon (Dunn et al 2012b).

This sequestration will continue for a certain length of time until an SOC equilibrium is reached. Equilibrium seems to occur after about 100 years in the case of switchgrass (Andress 2002) and 50 years in the case of miscanthus (Hill et al 2009, Scown et al 2012). This time-dependence of GHG emissions associated with LUC presents a challenge in biofuel LCA. The most appropriate time horizon for SOC changes and the treatment of future emissions as compared to near-term emissions is an open research question (Kløverpris and Mueller 2012, O'Hare et al 2009). On one hand, a near-term approach in which the time frame is two or three decades could be used. The advantages of this approach include assigning more importance to near-term events that are more certain. Some LCA standards, such as PAS 2050 (BSI 2011) advocate a 100 year time horizon for the LCA of any product. If such an extended time horizon is used, however, future emissions should be discounted, although the methodology for this discounting is unresolved. In addition, the uncertainty associated with land use for over a century is very large. Given these factors, we assume a 30 year period for both soil carbon modeling and for amortizing total LUC GHG emissions over biofuel production volume during this period. This approach, which aligns with the EPA's LCA methodology for the RFS (EPA 2010), may result in a slightly conservative estimate for the soil carbon sequestration that might be associated with switchgrass and miscanthus production, because lands producing these crops will continue to sequester carbon after the 30 year time horizon of this analysis. On the other hand, this selection gives a higher GHG sequestration rate per unit of biofuel since the total biofuel volume for amortization is smaller.

Our modeling with CCLUB indicates that of the feedstocks examined, corn ethanol had the largest LUC GHG emissions (9.1 g CO_2e MJ^{-1} of ethanol), whereas LUC emissions associated with miscanthus ethanol production caused substantial carbon sequestration (-12 g CO_2e MJ^{-1}). Switchgrass ethanol production results in a small amount of LUC emissions: 1.3 g CO_2e MJ^{-1}. LUC emissions associated with corn stover ethanol production result in a GHG sequestration of -1.2 g CO_2e MJ^{-1}. It is important to note that these results were generated by using one configuration of modeling assumptions in CCLUB. Elsewhere we describe how these results vary with alternative CCLUB configurations (Dunn et al 2012b).

TABLE 3: Cellulosic ethanol production parametric assumptions (per dry tonne of cellulosic biomass, except as noted).

Parameter: unit	Mean	P10	P90	Distribution function type
Corn stover collection				
Energy use for collection: MJ	219	197	241	Normal[a]
Supplemental N fertilizer: g	8488	6499	10476	Normal[a]
Supplemental P fertilizer: g	2205	1102	3307	Normal[a]
Supplemental K fertilizer: g	13228	7491	18964	Normal[a]
Switchgrass farming				
Farming energy use: MJ	144	89.1	199	Normal[b]
N fertilizer use: g	7716	4783	10649	Normal[b]
P fertilizer use: g	110	77	143	Normal[b]
K fertilizer use: g	220	154	287	Normal[b]
N_2O conversion rate of N fertilizer: %	1.525	0.413	2.956	Weibull[c]
Miscanthus farming				
Farming energy use: MJ	153	138	168	Normal[d]
N fertilizer use: g	3877	2921	4832	Normal[d]
P fertilizer use: g	1354	726	1981	Normal[d]
K fertilizer use: g	5520	3832	7209	Normal[d]
N_2O conversion rate of N fertilizer: %	1.525	0.413	2.956	Weibull[c]
Cellulosic ethanol productione				
Ethanol yield: l	375	328	423	Normal[f]
Electricity yield: kWh	226	162	290	Triangular[f]
Enzyme use: grams/kg of substrate (dry matter basis)	15.5	9.6	23	Triangular[g]
Yeast use: grams/kg of substrate (dry matter basis)	2.49	2.24	27.4	Normal[g]

[a]*By maximization of goodness-of-fit to the data compiled in Han et al (2011).* [b]*By maximization of goodness-of-fit to the data compiled in Dunn et al (2011).* [c]*Based on our new assessment of the literature, see supporting information (available at stacks.iop. org/ERL/7/045905/mmedia) for details.* [d]*By maximization of goodness-of-fit to the data compiled in Wang et al (2012).* [e]*Although we anticipated differences in plant yields and inputs among the three cellulosic feedstocks, we did not find enough data to quantify the differences for this study.* [f]*The type and shape of distribution functions were developed in Brinkman et al (2005). The means of the distributions were scaled later to the values in Wang et al (2011a).* gBy maximization of goodness-of-fit to the data compiled in Dunn et al (2012a).*

We have not conducted LUC GHG modeling for sugarcane ethanol. The EPA reported LUC GHG emissions for sugarcane ethanol of 5 g CO_2e MJ^{-1} (EPA 2010). This value does not include indirect effects of LUC beyond SOC changes, such as changes in emissions from rice fields and livestock production. The United Kingdom Department of Transport (E4Tech 2010) estimated indirect land use change (iLUC) associated with sugarcane ethanol as ranging between 18 and 27 g CO_2e MJ^{-1}. Another recent report estimates sugarcane LUC GHG emissions as 13 g CO_2e MJ^{-1} (ATLASS Consortium 2011). CARB estimated that these emissions were 46 g CO_2e MJ^{-1} (Khatiwada et al 2012) but is revisiting that value. The EU is proposing LUC GHG emissions of 13 g CO_2e MJ^{-1} (EC 2012). Without considering the CARB value, we decided to use LUC GHG emissions of 16 g CO_2e MJ^{-1} for sugarcane ethanol.

10.2.5 PETROLEUM GASOLINE

We made petroleum gasoline the baseline fuel to which the five ethanol types are compared. The emissions and energy efficiency associated with gasoline production are affected by the crude oil quality, petroleum refinery configuration, and gasoline quality. Of the crude types fed to US refineries, the Energy Information Administration (EIA 2012) predicts that in 2015 (the year modeled for this study), 13.4% of US crude will be Canadian oil sands. Based on EIA reports, we estimated 5.1% of US crude would be Venezuelan heavy and sour crude, and the remaining 81.5% would be conventional crude. The former two are very energy-intensive and emissions-intensive to recover and refine. US petroleum refineries are configured to produce gasoline and diesel with a two-to-one ratio by volume, while European refineries are with a one-to-two ratio. A gasoline-specific refining energy efficiency is needed for gasoline WTW analysis, and it is often calculated with several allocation methods (Wang et al 2004, Bredeson et al 2010, Palou-Rivera et al 2011). Also, methane flaring and venting could be a significant GHG emission source for petroleum gasoline. Table 4 lists the key parametric assumptions for petroleum gasoline.

TABLE 4: Petroleum gasoline parametric assumptions (per GJ of crude oil, except as noted).

Parameter: unit	Mean	P10	P90	Distribution function type
Conventional crude				
Conventional crude recovery efficiency: %	98.0	97.4	98.6	Triangular[a]
Heavy and sour crude recovery efficiency: %	87.9	87.3	88.5	Triangular[b]
CH_4 venting: g	7.87	6.26	9.48	Normal[c]
CO_2 from associated gas flaring/venting: g	1355	1084	1627	Normal[c]
Oil sands—surface mining (48% in 2015)				
Bitumen recovery efficiency: %	95.0	94.4	95.6	Triangular[d]
CH_4 venting: g	12.8	7.42	198	Normal[e]
CO_2 from associated gas flaring: g	187	83.9	289	Normal[e]
Hydrogen use for upgrade: MJ	84.2	67.4	101	Normal[d]
Oil sands—in situ production (52% in 2015)				
Bitumen recovery efficiency: %	85.0	83.6	86.5	Triangular[d]
Hydrogen use for upgrade: MJ	32.3	25.9	38.8	Normal[d]
Crude refining				
Gasoline refining efficiency: %	90.6	88.9	92.3	Normal[f]

[a]From Brinkman et al (2005). [b]Based on Rosenfeld et al (2009). [c]By maximization of goodness-of-fit to the data compiled in Palou-Rivera et al (2011). [d]From Larsen et al (2005). [e]Based on Bergerson et al (2012). [f]The type and shape of distribution functions were developed in Brinkman et al (2005). The means of the distributions were scaled later to the values in Palou-Rivera et al (2011).

10.2.6 TREATMENT OF CO-PRODUCTS IN BIOETHANOL AND GASOLINE LCA

Table 5 lists co-products, the products they displace and the co-product allocation methodologies for the six pathways included in this article. The displacement method is recommended by the International Standard Organization and was used by EPA and CARB. However, the energy allocation method was used by the European Commission. Wang et al (2011b) argued that while there is no universally accepted method to treat co-products in biofuel LCA, the transparency of methodology and the impacts of methodology choices should be presented in individual studies to better inform readers.

TABLE 5: Co-products of bioethanol and gasoline pathways and co-product allocation methodologies.

Pathway	Co-product	Displaced products	LCA method used in this study	Alternative LCA methods available in GREET	References
Corn ethanol	DGS[a]	Soybean, corn, and other animal feeds	Displacement	Allocation based on market revenue, mass or energy	Wang et al (2011b); Arora et al (2011)
Sugarcane ethanol	Electricity from bagasse	Conventional electricity	Allocation based on energy[b]	Displacement[c]	Wang et al (2008)
Cellulosic ethanol (corn stover, switchgrass and miscanthus)	Electricity from lignin	Conventional electricity	Displacement[d]	Allocation based on energy	Wang et al (2011b)
Petroleum gasoline	Other petroleum products	Other petroleum products	Allocation based on energy	Allocation based on mass, market revenue and process energy use	Wang et al (2004); Bredeson et al (2010); Palou-Rivera et al (2011)

[a]Dry mill corn ethanol plants produce dry and wet DGS with shares of 65% and 35% (on a dry matter basis), respectively. We include these shares in our analysis. [b]Electricity output accounts for 14% of the total energy output of sugarcane ethanol plants. With such a significant share of electricity, we decided to use the energy allocation method for ethanol and electricity rather than the displacement method. [c]With the displacement method, if we assume that the co-produced electricity displaces the Brazilian average electricity mix (with 83% from hydro power), the sugarcane ethanol results are similar to those when the energy allocation method is used. If the co-produced electricity displaces NG combined cycle power, WTW sugarcane ethanol GHG emissions are reduced by 21 g CO_2e MJ^{-1}. [d]We assumed that co-produced electricity replaces the US average electricity mix in 2015 (with 44% from coal and 21% from NG (EIA 2012) and a GHG emission rate of 635 g CO_2e kWh^{-1}). If co-produced electricity displaces the US midwest generation mix (with 74% from coal and 4% from NG and a GHG emission rate of 844 g CO_2e kWh^{-1}), cellulosic ethanol WTW GHG emissions are reduced by 5.7 g CO_2e MJ^{-1}. If co-produced electricity displaces NG combined cycle power (with a GHG emission rate of 539 g CO^2e kWh^{-1}), cellulosic ethanol GHG emissions are increased by 2.5 g CO_2e MJ^{-1} from the base case.

Figure 3 from Michael Wang et al 2012 Environ. Res. Lett. 7 045905

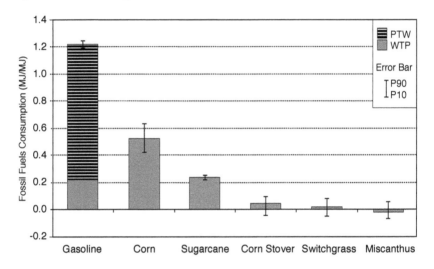

FIGURE 3: Well-to-wheels results for fossil energy use of gasoline and bioethanol.

10.3 RESULTS

We present WTW results for energy use and GHG emissions for the five bioethanol pathways and baseline gasoline (a blending stock without ethanol or other oxygenates). Energy use results for this study include total energy use, fossil energy use, petroleum use, natural gas use and coal use. Because of space limitations, only fossil energy use results (including petroleum, coal and natural gas) are presented here. GHG emissions here are CO_2-equivalent emissions of CO_2, CH_4 and N_2O, with 100 year global warming potentials of 1, 25 and 298, respectively, per the recommendation of the International Panel on Climate Change (Eggleston et al 2006).

Figure 3 presents WTW results for fossil energy use per MJ of fuel produced and used. The chart presents the well-to-pump (WTP) stage (more precisely, in the bioethanol cases, field-to-pump stage) and pump-to-wheels (PTW) stage. The WTP and PTW bars together represent WTW results. The error bars represent values with P10 (the lower end of the line) and P90 (the higher end of the line) for WTW results.

Selection of the MJ functional unit here means that energy efficiency differences between gasoline and ethanol vehicles are not taken into account. On an energy basis (or gasoline-equivalent basis), vehicle efficiency differences for low-level and mid-level blends of ethanol in gasoline are usually small. If engines are designed to take advantage of the high octane number of ethanol, however, high-level ethanol blends could improve vehicle efficiency.

TABLE 6: Energy balance and energy ratio of bioethanol.

	Corn	Sugarcane	Corn stover	Switchgrass	Miscanthus
Energy balance (MJ l^{-1})^a	10.1	16.4	20.4	21.0	21.4
Energy ratio	1.61	4.32	4.77	5.44	6.01

^{a}A liter of ethanol contains 21.3 MJ of energy (lower heating value)

For petroleum gasoline, the largest amount of fossil energy is used in the PTW stage because gasoline energy is indeed fossil-based. In contrast, the five ethanol pathways do not consume fossil energy in the PTW stage. With regard to WTP fossil energy use, corn ethanol has the largest amount due to the intensive use of fertilizer in farming and use of energy (primarily NG) in corn ethanol plants. Other ethanol pathways have minimum fossil energy use. In fact, the P10 fossil energy values for the three cellulosic ethanol types are negative for two reasons. First, fossil energy use during farming and ethanol production for these pathways is minimal. Second, the electricity generated in cellulosic ethanol plants can displace conventional electricity generation, which, in the US, is primarily fossil energy based. Relative to gasoline, ethanol from corn, sugarcane, corn stover, switchgrass and miscanthus, on average, can reduce WTW fossil energy use by 57%, 81%, 96%, 99% and 100%, respectively.

An energy balance or energy ratio is often presented for bioethanol to measure its energy intensity. Table 6 presents energy balances and ratios of the five bioethanol pathways. The energy balance is calculated as the difference between the energy content of ethanol and the fossil energy used to produce it. Energy ratios are calculated as the ratio between the two. All five ethanol types have positive energy balance values and energy ratios greater than one.

Figure 4 shows WTW GHG emissions of the six pathways. GHG emissions are separated into WTP, PTW, biogenic CO_2 (i.e., carbon in bioethanol) and LUC GHG emissions. Combustion emissions are the most significant GHG emission source for all fuel pathways. However, in the five bioethanol cases, biogenic CO_2 in ethanol offsets ethanol combustion GHG emissions almost entirely. LUC GHG emissions, as discussed in an earlier section, are from the CCLUB simulations for the four bioethanol pathways (corn, corn stover, switchgrass and miscanthus). LUC emissions of Brazilian sugarcane ethanol are based on our review of available literature. It is not possible to maintain a consistent analytical approach among these unharmonized literature studies of sugarcane ethanol and between them and CCLUB modeling results. Because of the ongoing debate regarding the values and associated uncertainties of LUC GHG emissions, we provide two separate sets of results for ethanol: one with LUC emissions included, and the other with LUC emissions excluded.

TABLE 7: WTW GHG emission reductions for five ethanol pathways (relative to WTW GHG emissions for petroleum gasoline). (Note: Values in the table are GHG reductions for P10–P90 (P50), all relative to the P50 value of gasoline GHG emissions.)

WTW GHG emission reductions	Corn	Sugarcane	Corn stover	Switchgrass	Miscanthus
Including LUC emissions	19–48% (34%)	40–62% (51%)	90–103% (96%)	77–97% (88%)	101–115% (108%)
Excluding LUC emissions	29–57% (44%)	66–71% (68%)	89–102% (94%)	79–98% (89%)	88–102% (95%)

Of the five bioethanol pathways, corn and sugarcane ethanol have significant WTP GHG emissions and LUC GHG emissions. Miscanthus ethanol has significant negative LUC GHG emissions due to the increased

SOC content from miscanthus growth. Sugarcane ethanol shows great variation in LUC emissions, mainly due to differences in assumptions and modeling methodologies among the reviewed studies. Table 7 shows numerical GHG emission reductions of the five ethanol pathways relative to those of petroleum gasoline.

The pie charts in figure 5 show contributions of key life-cycle stages to WTW GHG emissions for the six pathways. With regard to gasoline WTW GHG emissions, 79% are from combustion of gasoline and 12% are from petroleum refining. Crude recovery and transportation activities contribute the remaining 9%. For corn ethanol, ethanol plants account for 41% of total GHG emissions; fertilizer production and N_2O emissions from cornfields account for 36%; LUC accounts for 12%; and corn farming energy use and transportation activities account for small shares. For sugarcane ethanol, LUC accounts for 36% of total GHG emissions (however, LUC GHG emissions data here are from a literature review rather than our own modeling). Transportation of sugarcane and ethanol contributes to 24% of total GHG emissions. Together, fertilizer production and N_2O emissions from sugarcane fields account for 20% of these emissions. Finally, the contribution of sugarcane farming to WTW GHG emissions is 11%.

Although for corn ethanol, the greatest contributor to life-cycle GHG emissions is the production of ethanol itself, this step is less significant in the life cycle of sugarcane ethanol because sugar mills use bagasse to generate steam and electricity. Another contrast between these two sugar-derived biofuels is the transportation and distribution (T&D) stage. Corn ethanol, produced domestically in the US, is substantially less affected by T&D than is sugarcane ethanol, which is trucked for long distances to Brazilian ports and transported across the ocean via ocean tankers to reach US consumers.

For the three cellulosic ethanol pathways, ethanol production is the largest GHG emission source. Fertilizer production and associated N_2O emissions (only in the case of switchgrass and miscanthus) are the next largest GHG emission source. Farming and transportation activities also have significant emission shares. One notable aspect of figure 5(e) is the positive contribution of LUC GHG emissions in the switchgrass ethanol life cycle when compared to the other cellulosic feedstocks, which may sequester GHG as a result of LUC. These results are explained elsewhere (Dunn et al 2012b).

Figure 4 from Michael Wang et al 2012 Environ. Res. Lett. 7 045905

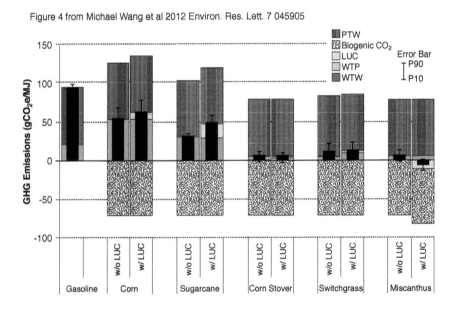

FIGURE 4: Well-to-wheels results for greenhouse gas emissions in CO$_2$e for six pathways.

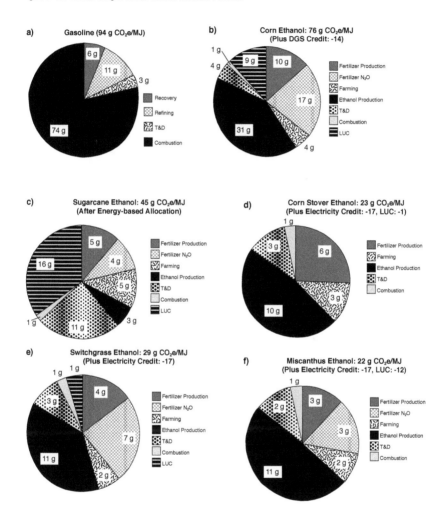

FIGURE 5: Shares of GHG emissions by activities for (a) gasoline, (b) corn ethanol, (c) sugarcane ethanol, (d) corn stover ethanol, (e) switchgrass ethanol and (f) miscanthus ethanol (results were generated by using the co-product allocation methodologies listed in table 6).

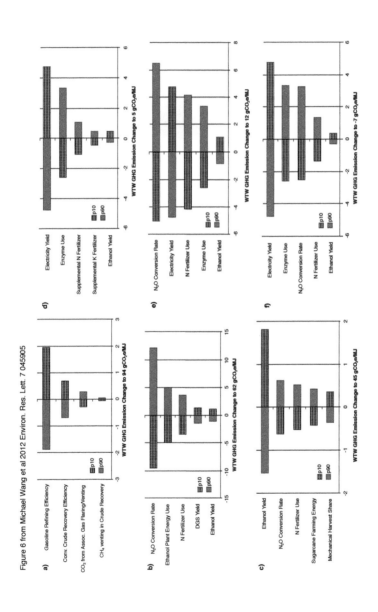

FIGURE 6: Sensitivity analysis results for (a) conventional crude to gasoline, (b) corn ethanol, (c) sugarcane ethanol, (d) corn stover ethanol, (e) switchgrass ethanol and (f) miscanthus ethanol.

To show the importance of key parameters affecting WTW GHG emissions results for a given fuel pathway, we conducted a sensitivity analysis of GHG emissions with GREET for all six pathways with P10 and P90 values as the minimum and maximum value for each parameter. We present the five most influential parameters for each pathway in the so-called tornado charts in figure 6.

For petroleum gasoline, the gasoline refining efficiency and recovery efficiency of the petroleum feedstock are the most sensitive parameters. For corn ethanol, the N_2O conversion rate in cornfields is the most sensitive factor, followed by the ethanol plant energy consumption. Enzyme and yeast used in the corn ethanol production process are not among the five most influential parameters in the corn ethanol life cycle. For sugarcane ethanol, the most significant parameters, in order of importance, are ethanol yield per unit of sugarcane, the N_2O conversion rate in sugarcane fields, nitrogen fertilizer usage intensity, sugarcane farming energy use and the mechanical harvest share. Sugarcane farming is evolving as mechanical harvesting becomes more widespread and mill by-products are applied as soil amendments. We thus expect to see shifts in the identity and magnitude of influence of the key parameters in the sugarcane-to-ethanol pathway in the future.

The three cellulosic ethanol pathways have similar results. The electricity credit is the most significant parameter (except for switchgrass ethanol, for which the N_2O conversion rate is the most significant). Enzyme use is a more significant factor in cellulosic ethanol pathways than in the corn ethanol pathway because the greater recalcitrance of the feedstock currently requires higher enzyme dosages in the pretreatment stage (Dunn et al 2012a). The impact of fertilizer-related parameters on WTW GHG emissions results depends, as one would expect, on the fertilizer intensity of feedstock farming (see table 3).

The strong dependence of results on the N_2O conversion rate is notable for four out of the five ethanol pathways (the exception is corn stover, where the same amount of nitrogen in either in the stover or supplemental fertilizer results in same amount of N_2O emissions, with or without stover collection). Great uncertainty exists regarding N_2O conversion rates in agricultural fields because many factors (including soil type, climate, type of fertilizer and fertilizer application method) affect the conversion. We conducted an extensive literature review for this study to revise N_2O conversion rates in GREET (see supporting information available at stacks. iop.org/ERL/7/045905/mmedia). The original GREET conversion rate was based primarily on IPCC tier 1 rates. With newly available data, we adjusted our direct conversion rates in cornfields upward (see supporting information available at stacks.iop.org/ERL/7/045905/mmedia for details). In particular, we developed a Weibull distribution function for direct and indirect N_2O emissions together with a mean value of 1.525%, a P10 value of 0.413% and P90 value of 2.956%. In comparison, our original distribution function for total N_2O conversion rates was a triangular distribution, with a most likely value of 1.325%, a minimum value of 0.4% and a maximum value of 2.95%.

10.4 DISCUSSION

Our results for cellulosic ethanol are in line with two recent studies that reported life-cycle GHG emissions of switchgrass and miscanthus ethanol. Monti et al (2012) reported that switchgrass ethanol life-cycle GHG emissions are 63% to 118% lower than gasoline, based on a literature review. Scown et al (2012) conducted an LCA of miscanthus ethanol and reported its life-cycle GHG emissions as being -26 g CO_2e MJ^{-1} of ethanol when impacts of both co-produced electricity and soil carbon sequestration were included. We estimate slightly lower reductions for sugarcane ethanol than did Seabra et al (2011) and Macedo et al (2008). Our results for corn ethanol, however, contrast with those of Searchinger et al (2008) and Hill et al (2009), who predicted that corn ethanol would have a greater life-cycle GHG impact than gasoline, mainly due to LUC GHG emissions among those studies and ours.

Advances and complexities in ethanol production technologies, especially for cellulosic ethanol, could alter bioethanol LCA results in the future. For example, although we examined corn and cellulosic ethanol plants separately in this article, when cellulosic ethanol conversion technologies become cost competitive, it is conceivable that cellulosic feedstocks could be integrated into existing corn ethanol plants, with appropriate modifications. Thus, an integrated system with both corn and cellulosic feedstocks (especially corn stover) could be evaluated. Such an integrated ethanol plant might have some unique advantages if one feedstock suffered from decreased production (e.g., the anticipated reduction in corn production in key Midwestern states in 2012 as a result of the severe drought).

In addition, cellulosic ethanol plants and their ethanol yields could be significantly different among different feedstocks. The source of the energy intensity data for converting a cellulosic feedstock to ethanol via a biochemical conversion process that we used in our WTW simulations was with the process of converting corn stover (Humbird et al 2011). We did not obtain separate conversion energy intensity data for other cellulosic feedstocks. In the future, we will examine the differences in both ethanol yield and co-produced electricity among different cellulosic feedstocks.

Co-produced electricity is another significant yet uncertain factor contributing to cellulosic ethanol's GHG benefits. Electricity yields in cellulosic ethanol plants, however, are highly uncertain. In fact, it is not entirely certain that cellulosic ethanol plants will install capital-intensive CHP equipment that would permit the export of electricity to the grid.

Considering the feedstock production phase, the significant difference in WTW results between switchgrass and miscanthus ethanol is caused mainly by the large difference in yield between the two crops (12 tonnes ha^{-1} for switchgrass versus 20 tonnes ha^{-1} for miscanthus). The high yield of miscanthus results in a significant increase in SOC content in simulations that use the CENTURY model (Kwon et al 2012), which is based on the common understanding that a high biomass yield can result in high below ground biomass accumulation. This implies that any cellulosic feedstock with a high yield, such as miscanthus, could sequester significant amounts of GHGs. Thus, instead of interpreting the results presented here as unique to switchgrass and miscanthus, we suggest that the results can indicate the differences between high-yield and low-yield dedicated energy crops.

For all bioethanol pathways, the strong dependence of GHG emission results on the N_2O conversion rate of N fertilizer suggests the need to continuously improve the efficiency with which N fertilizer is used in farm fields and the need to estimate that parameter more precisely. The needs are especially important with regard to nitrogen dynamics in sugarcane fields and cornfields.

In addition, the seasonal harvest of cellulosic feedstocks to serve the annual operation of cellulosic ethanol plants requires the long-time storage of those feedstocks. Feedstock loss during storage as well as during harvest and transportation is an active research topic. We will include cellulosic feedstock loss in our future WTW analysis of cellulosic ethanol pathways.

The WTW GHG emissions of petroleum gasoline are also subject to significant uncertainties. Some researchers estimated GHG emissions associated with indirect effects from petroleum use, such as those from military operations in the Middle East (Liska and Perrin 2010). Depending on the ways that GHG emissions from military operations are allocated, those emissions could range from 0.9 to 2.1 g MJ^{-1} of gasoline (Wang et al 2011a). Moreover, GHG emissions associated with oil recovery can vary considerably, depending on the type of recovery methods used, well depth, and flaring and venting of CH_4 emissions during recovery (Rosenfeld et al 2009, Brandt 2012).

10.5 CONCLUSIONS

Bioethanol is the biofuel that is produced and consumed the most globally. The US is the dominant producer of corn-based ethanol, and Brazil is the dominant producer of sugarcane-based ethanol. Advances in technology and the resulting improved productivity in corn and sugarcane farming and ethanol conversion, together with biofuel policies, have contributed to the significantly expanded production of both types of ethanol in the past 20 years. These advances and improvements have helped bioethanol achieve increased energy and GHG emission benefits when compared with those of petroleum gasoline.

We used an updated, upgraded version of the GREET model to estimate life-cycle energy consumption and GHG emissions for five bioethanol production pathways on a consistent basis. Even when we included highly debated LUC GHG emissions, when the feedstock was changed from corn to sugarcane and then to cellulosic biomass, bioethanol's reductions in energy use and GHG emissions, when compared with those of gasoline, increased significantly. Thus, in the long term, the cellulosic ethanol production options will offer the greatest energy and GHG emission benefits. Policies and research and development efforts are in place to promote such a long-term transition.

REFERENCES

1. Andress D 2002 Soil Carbon Changes for Bioenergy Crops (report prepared for Argonne National Laboratory and US Department of Energy) (http://greet.es.anl.gov/publication-rfihxb2h, accessed 26 October 2012)
2. Argonne National Laboratory 2012 GREET Model (http://greet.es.anl.gov/)
3. Arora S, Wu M and Wang M 2011 Update of Distillers Grains Displacement Ratios for Corn Ethanol Life-Cycle Analysis Argonne National Laboratory Report ANL/ESD/11-1
4. ATLASS Consortium 2011 Assessing the Land Use Change Consequences of European Biofuel Policies (provided to the Directorate General for Trade of the European Commission) (http://trade.ec.europa.eu/doclib/docs/2011/october/tradoc_148289.pdf, accessed 8 August 2012)
5. Bergerson J A, Kofoworola O, Charpentier A D, Sleep S and MacLean H L 2012 Life cycle greenhouse gas emissions of current oil sands technologies: surface mining and in situ applications Environ. Sci. Technol. 46 7865–74
6. Borrion A L, McManus M C and Hammond G P 2012 Environmental life cycle assessment of lignocellulosic conversion to ethanol: a review Renew. Sustain. Energy Rev. 16 4638–50
7. Braga do Carmo J et al 2012 Infield greenhouse gas emissions from sugarcane soils in Brazil: effects from synthetic and organic fertilizer application and crop trash accumulation GCB Bioenergy at press (doi:10.1111/j.1757-1707.2012.01199.x)
8. Brandt A 2012 Variability and uncertainty in life cycle assessment models for greenhouse gas emissions from Canadian oil sands production Environ. Sci. Technol. 43 1253–64
9. Bredeson L, Quiceno-Gonzalez R and Riera-Palou X 2010 Factors driving refinery CO2 intensity, with allocation into products Int. J. Life Cycle Assess. 15 817–27
10. Brinkman N, Wang M, Weber T and Darlington T 2005 Well-to-Wheels Analysis of Advanced Fuel/Vehicle Systems—A North American Study of Energy Use, Greenhouse Gas Emissions, and Criteria Pollutant Emissions (Argonne, IL: Argonne National Laboratory)
11. BSI 2011 PAS 2050: 2011 Specification for the Assessment of the Life Cycle Greenhouse Gas Emissions of Goods and Services (London: British Standards)
12. Burnham A, Han J, Clark C E, Wang M Q, Dunn J B and Palou-Rivera I 2012 Life-cycle greenhouse gas emissions of shale gas, natural gas, coal, and petroleum Environ. Sci. Technol. 46 619–27
13. CARB (California Air Resources Board) 2009 Proposed Regulation for Implementing Low Carbon Fuel Standards (Staff Report: Initial Statement of Reasons vol

1) (Sacramento, CA: California Environmental Protection Agency, Air Resources Board) (www.arb.ca.gov/regact/2009/lcfs09/lcfsisor1.pdf)

14. Chum H et al 2011 Bioenergy IPCC Special Report on Renewable Energy Sources and Climate Change Mitigation ed O Edenhofer (Cambridge: Cambridge University Press)

15. DOE (US Department of Energy) 2011 US Billion-Ton Update: Biomass Supply for a Bioenergy and Bioproducts Industry (Washington, DC: Oak Ridge National Laboratory for DOE Office of Energy Efficiency and Renewable Energy, Biomass Program)

16. Dunn J B, Eason J and Wang M Q 2011 Updated Sugarcane and Switchgrass Parameters in the GREET Model (http://greet.es.anl.gov/publication-updated_sugarcane_switchgrass_params)

17. Dunn J B, Mueller S, Wang M Q and Han J 2012a Energy consumption and greenhouse gas emissions from enzyme and yeast manufacture for corn and cellulosic ethanol production Biotechnol. Lett. 34 2259–63

18. Dunn J B, Mueller S and Wang M Q 2012b Land-use change and greenhouse gas emissions from corn and cellulosic ethanol Biotechnol. Biofuels submitted

19. E4Tech 2010 A Causal Descriptive Approach to Modeling the GHG Emissions Associated with the Indirect Land Use Impacts of Biofuels (provided to the UK Department of Transport) (www.e4tech.com/en/overview-publications.cfm, accessed 8 August 2012)

20. EC (European Commission) 2012 Proposal for a Directive of the European Parliament and of the Council of Biofuel Land Use Change Emissions (Brussels: EC)

21. Edgerton M D et al 2010 Commercial scale corn stover harvests using field-specific erosion and soil organic matter targets Sustainable Alternative Fuel Feedstock Opportunities, Challenges, and Roadmaps for Six US Regions (Proc. Sustainable Feedstocks for Advanced Biofuels Workshop) ed R Braun, D Karlen and D Johnson (Ankeny, IA: Soil and Water Conservation Society) pp 247–56

22. Eggleston S L, Buendia L, Miwa K, Ngara T and Tanabe K 2006 2006 IPCC Guidelines for National Greenhouse Gas Inventories (General Guidance and Reporting vol 1) (Hayama: Institute for Global Environmental Strategies)

23. EIA (Energy Information Administration) 2012 Annual Energy Outlook 2012 (Washington, DC: US Department of Energy) (www.eia.gov/forecasts/aeo/pdf/0383(2012).pdf, accessed 20 July 2012)

24. EPA (US Environmental Protection Agency) 2010 Renewable Fuel Standard Program (RFS2) Regulatory Impact Analysis (Washington, DC: US Environmental Protection Agency)

25. Fargione J, Hill J, Tilman D, Polasky S and Hawthorne P 2008 Land clearing and the biofuel carbon debt Science 319 1235–3

26. Farrell A E, Plevin R J, Turner B T, Jones A D, O'Hare M and Kammen D M 2006 Ethanol can contribute to energy and environmental goals Science 311 506–8

27. Han J, Elgowainy A, Palou-Rivera I, Dunn J B and Wang M Q 2011 Well-to-Wheels Analysis of Fast Pyrolysis Pathways with GREET Argonne National Laboratory Report ANL/ESD/11-8

28. Hill J, Polasky S, Nelson E, Tilman D, Huo H, Ludwig L, Neumann J, Zheng H and Bonta D 2009 Climate change and health costs of air emissions from biofuels and gasoline Proc. Natl Acad. Sci. 106 2077–82

29. Humbird D et al 2011 Process Design and Economics for Biochemical Conversion of Lignocellulosic Biomass to Ethanol National Renewable Energy Laboratory Report NREL/TP-5100-47764

30. IEA (International Energy Agency) 2012 Energy Technology Perspective 2012: Pathways to a Clean Energy System (Paris: International Energy Agency)

31. Khatiwada D, Seabra J, Silveira S and Walter A 2012 Accounting greenhouse gas emissions in the lifecycle of Brazilian sugarcane bioethanol: Methodological references in European and American regulations Energy Policy 47 384–97

32. Kløverpris J H and Mueller S 2012 Baseline time accounting: considering global land use dynamics when estimating the climate impact of indirect land use change caused by biofuels Int. J. Life Cycle Assess. at press (doi:10.1007/s11367-012-0488-6)

33. Kwon H, Wander M M, Mueller S and Dunn J B 2012 Modeling state-level soil carbon emissions factors under various scenarios for direct land use change associated with United States biofuel feedstock production Biomass Bioenergy at press

34. Larsen R, Wang M, Wu Y, Vyas A, Santini D and Mintz M 2005 Might Canadian oil sands promote hydrogen production for transportation? Greenhouse gas emission implications of oil sands recovery and upgrading World Resour. Rev. 17 220–42

35. Liska A J and Perrin R K 2010 Securing foreign oil: acase for including military operations in the climate change impact of fuels Environment 52 9–22

36. Liska A J et al 2009 Improvements in life cycle energy efficiency and greenhouse gas emissions of corn-ethanol J. Indust. Ecol. 13 58–74

37. Macedo I D C, Leal M R L V and Seabra J E A R 2004 Assessment of Greenhouse Gas Emissions in the Production and Use of Fuel Ethanol in Brazil (prepared for the state of Sao Paulo, Brazil) (www.wilsoncenter.org/sites/default/files/brazil.unicamp. macedo.greenhousegas.pdf)

38. Macedo I D C, Seabra J E A and Silva J E A R 2008 Greenhouse gases emissions in the production and use of ethanol from sugarcane in Brazil: the 2005/2006 averages and a prediction for 2020 Biomass Bioenergy 32 582–95

39. MacLean H and Spatari S 2009 The contribution of enzymes and process chemicals to the life cycle of ethanol Environ. Res. Lett. 4 014001

40. IOPscience Monti A, Lorenzo B, Zatta A and Zegada-Lizarazu W 2012 The contribution of switchgrass in reducing GHG emissions GCB Bioenergy 4 420–34

41. Mu D, Seager T, Rao P S and Zhao F 2010 Comparative life cycle assessment of lignocellulosic ethanol production: biochemical versus thermochemical conversion Environ. Manag. 46 565–78

42. Mueller S, Dunn J B and Wang M Q 2012 Carbon Calculator for Land Use Change from Biofuels Production (CCLUB) Users' Manual and Technical Documentation (Argonne, IL: Argonne National Laboratory) (http://greet.es.anl.gov/publication-cclub-manual)

43. Neeft J et al 2012 BioGrace—Harmonized Calculations of Biofuel Greenhouse Gas Emissions in Europe (www.biograce.net)

44. O'Hare M, Plevin R J, Martin J I, Jones A D, Kendall A and Hopson E 2009 Proper accounting for time increases crop-based biofuels' greenhouse gas deficit versus petroleum Environ. Res. Lett. 4 024001

45. IOPscience Palou-Rivera I, Han J and Wang M 2011 Updates to Petroleum Refining and Upstream Emissions (Argonne, IL: Argonne National Laboratory) (http://greet.es.anl.gov/publication-petroleum)

46. RFA (Renewable Fuels Association) 2012 2012 Ethanol Industry Outlook: Accelerating Industry Innovation (Washington, DC: Renewable Fuels Association)

47. Rosenfeld J, Pont J, Law L, Hirshfeld D and Kolb J 2009 Comparison of North American and Imported Crude Oil Life Cycle GHG Emissions (Calgary, AB: TIAX LLC and MathPro Inc. for Alberta Energy Research Institute) TIAX: Case No. D5595

48. Scown C D et al 2012 Lifecycle greenhouse gas implications of US national scenarios for cellulosic ethanol production Environ. Res. Lett. 7 014011

49. IOPscience Seabra J E A, Macedo I C, Chum H L, Faroni C E and Sarto C A 2011 Life cycle assessment of Brazilian sugarcane products: GHG emissions and energy use Biofuels, Bioprod. Biorefining 5 519–32

50. Searchinger T et al 2008 Use of US croplands for biofuels increases greenhouse gases through emissions from land use change Science 319 1238–40

51. Sheehan J, Aden A, Paustian K, Killian K, Brenner J, Walsh M and Nelsh R 2008 Energy and environmental aspects of using corn stover for fuel ethanol J. Ind. Ecol. 7 117–46

52. Sokhansanj S, Mani S, Turhollow A, Kumar A, Bransby D, Lynd L and Laser M 2009 Large-scale production, harvest and logistics of switchgrass (Panicumbirgatum L.)—current technology and envisioning a mature technology Biofuels, Bioprod. Biorefining 3 124–41

53. Somerville C, Young H, Taylor C, Davis S C and Long S P 2010 Feedstocks for lignocellulosic biofuels Science 329 791–2

54. Taheripour F, Tyner W E and Wang M Q 2011 Global Land Use Changes Due to the US Cellulosic Biofuel Program Simulated with the GTAP Model (Argonne, IL: Argonne National Laboratory) (http://greet.es.anl.gov/publication-luc_ethanol)

55. Tyner W, Taheripour F, Zhuang Q, Birur D and Baldos U 2010 Land Use Changes and Consequent CO2 Emissions due to US Corn Ethanol Production: A Comprehensive Analysis (West Lafayette, IN: Department of Agricultural Economics, Purdue University)

56. UNICA (Brazilian Sugarcane Association) 2012 UNICA Data Center (www.unica-data.com.br/index.php?idioma=2, accessed 16 July 2012)

57. Van Deusen P C and Heath L S 2010 Weighted analysis methods for mapped plot forest inventory data: tables, regressions, maps and graphs Forest Ecol. Manage. 260 1607–12

58. Wang M, Han J, Haq Z, Tyner W, Wu M and Elgowainy A 2011a Energy and greenhouse gas emission effects of corn and cellulosic ethanol with technology improvements and land use changes Biomass Bioenergy 35 1885–96

59. Wang M, Huo H and Arora S 2011b Methodologies of dealing with co-products of biofuels in life-cycle analysis and consequent results within the US context Energy Policy 539 5726–36

60. Wang M, Lee H and Molburg J 2004 Allocation of energy use and emissions to petroleum refining products: implications for life-cycle assessment of petroleum transportation fuels Int. J. Life Cycle Assess. 9 34–44

61. Wang M, Wu M and Huo H 2007 Life-cycle energy and greenhouse gas emission impacts of different corn ethanol plant types Environ. Res. Lett. 2 024001

62. IOPscience Wang M, Wu M, Huo H and Liu J 2008 Life-cycle energy use and greenhouse gas emission implications of Brazilian sugarcane ethanol production simulated with the GREET model Int. Sugar J. 110 527–45

63. Wang Z, Dunn J B and Wang M Q 2012 GREET Model Miscanthus Parameter Development (Argonne, IL: Argonne National Laboratory) (http://greet.es.anl.gov/publication-micanthus-params)

64. Whitaker J, Ludley K E, Rowe R, Taylor G and Howard D C 2010 Sources of variability in greenhouse gas and energy balances for biofuel production: a systematic review GCB Bioenergy 2 99–112

CHAPTER 11

Lessons from First Generation Biofuels and Implications for the Sustainability Appraisal of Second Generation Biofuels

ALISON MOHR AND SUJATHA RAMAN

11.1 INTRODUCTION

The story of biofuels has been described as one of 'riches to rags' (Sengers et al., 2010). Initially cornucopian views of the potential of biofuels have been challenged under the weight of increasing speculation that their pace of development was racing ahead of understanding of the range of direct and indirect sustainability impacts of this technology. UK and EU targets for renewable fuels in the transport sector have further compounded perceptions of an unfettered dash for biofuels. Media headlines linking the rise of vast biofuel plantations in various parts of the world with rising food prices provoked a rapid shift in thinking about this technology in the second half of the 2000s. No longer is it possible to encounter the term 'energy crops' without some awareness of the potential conflict with the

Lessons from First Generation Biofuels and Implications for the Sustainability Appraisal of Second Generation Biofuels. © Mohr A and Raman S. Energy Policy 63 (2013), DOI: 10.1016/j.enpol.2013.08.033. Licensed under a Creative Commons Attribution 3.0 Unported License, http://creativecommons.org/licenses/by/3.0/.

use of agricultural land for food encapsulated by the term 'food vs. fuel'. Other social, environmental, economic and ethical challenges are emerging especially with respect to so-called 'first generation' biofuels produced from food crops.

Biofuels have been roughly classified to distinguish between first generation (1G) biofuels produced primarily from foods crops such as grains, sugar cane and vegetable oils and second generation (2G) biofuels produced from cellulosic energy crops such as miscanthus and SRC willow, agricultural forestry residues or co-products such as wheat straw and woody biomass. Opposition to 1G biofuels is generally assumed to be about conflict with food security. Second generation biofuels are widely seen as a sustainable response to the increasing controversy surrounding 1G, and thus distinct from it. Indeed, it has been suggested that 2G biofuels raise few ethical or sustainability issues (e.g., Charles et al., 2007 and Nuffield Council on Bioethics, 2011). But will the emergence of 2G biofuels dispel claims of 'food vs. fuel' conflicts and what new challenges might they raise? As the world's first commercial-scale cellulosic ethanol plant in Crescentino, Italy began operating at the end of 2012, this question is particularly timely.

11.2 AIMS AND METHODS

Examining the lessons arising from the controversy surrounding 1G biofuels, this paper assesses their relevance for perceptions of sustainability of 2G biofuels and considers the policy challenges for managing the transition to a sustainable UK bioenergy system, with particular emphasis on lignocellulosic options for biofuels. In doing so, we build on work suggesting that the ubiquitous reference to 'food vs. fuel' conflicts does not adequately capture the challenges posed by 1G biofuels (Raman and Mohr, in press). If this is the case, the case for 2G biofuels likewise needs to address a wider range of issues than conflict with food security alone. We draw on our experience as social scientists embedded in a major UK scientific programme on 2G biofuels where a key aspect of our work is to explore different stakeholder assessments of the sustainability of biofuels in the UK, in the context of a global bioenergy system.

TABLE 1: Implications for the sustainability appraisal of 2G biofuels of challenges arising from 1G.

Sustainability challenge	First generation (1G)	Second generation (2G)	Policy challenges
Food security	Negative impact on food security is the biggest concern raised about using food crops and oils for producing fuel ('food vs. fuel').	May be relevant for non-food energy crops if grown on land having value for food production, including lignocellulose sourced from the global South.	Bioenergy can help ameliorate food price rises linked to fuel and fertiliser price rises (Murphy et al., 2011).
	Precise role of 1G biofuels in food price spikes contested.	Use of agricultural residues would not constitute a direct conflict with food (but questions may arise where biofuel production using residues is linked with 1G feedstocks).	The efficacy of re-using marketable by-products and residues of bioethanol production relative to other methods for improving soil fertility should be considered (Singh et al., 2010).
		Residues such as straw may be part of the animal feed mix and thus indirectly linked to the food chain.	Use of land for grazing or animal feed could also be independently assessed (Wassenaar and Kay, 2008).
Large-scale land acquisition	Where land has been acquired for sourcing biomass from global South, violations have been reported of people's rights and livelihoods (Friends of the Earth, 2007 and Oxfam, 2007).	Land officially designated as 'marginal' or degraded, but suitable for 2G feedstocks, may be relied on by the poor for subsistence.	Sustainability of land use is a 'wicked' problem that extends beyond debates on the sustainability of agriculture to raise broader questions about land use and management practices as a whole.
			Reliable monitoring of land acquisition in developing countries may be difficult given major disparities between large companies and local people (van der Horst and Vermeylen, 2011).

TABLE 1: *Cont.*

Sustainability challenge	First generation (1G)	Second generation (2G)	Policy challenges
	LUC of natural habitats and other ecologically valuable land acquired for biofuel plantations has been linked to loss of bio-diversity (Pilgrim and Harvey, 2010; civil society stakeholder interview, 6 December 2011).	For agricultural residues, the sustainable use of land in which the crop is grown may be relevant. Biofuel companies may target higher quality agricultural land (science stakeholder interview, 26 October 2011; civil society stakeholder interview, 20 December 2011) or forested land that will provide additional income from logging (van der Horst and Vermeylen, 2011), and intensification can further exacerbate ecological impacts.	Relative priorities for different uses of land may differ according to different communal or cultural value sets (Thornley et al., 2009), or the uneven distribution of climate change impacts, and need to be considered in all parts of the world.
GHG balance	iLUC effects including the release of carbon stocks from conversion of forests, peatlands or grasslands for biofuel crops (Fargione et al., 2008) will reduce carbon savings. Impact of nitrogen fertilizers and energy costs of transporting feedstocks can affect net energy balance.	If feedstocks were sourced by felling of forests, or where the use of 'marginal' land which is in fact a source of food stimulates iLUC effects, GHG balance may be questioned. Soil organic carbon content is affected by removal of straw (Thornley et al., 2009).	Given complexity of iLUC and its relevance to agriculture as a whole, deforestation may be more reliably addressed through dedicated policies rather than inclusion in GHG calculations for biofuels (Zilberman and Hochman, 2010). Results of calculations depend on system boundaries and assumptions which need to be explicit to avoid misuse by decision-makers (Singh et al., 2010; Modelling Uncertainties Workshop, 2012).

TABLE 1: *Cont.*

Sustainability challenge	First generation (1G)	Second generation (2G)	Policy challenges
	Most conventional biofuels depend on fossil fuels for their production. (House of Commons Environmental Audit Committee, 2008).	Net energy balance is relevant given the energy inputs needed to break down ligno-cellulosic material and for transportation of bulky residues.	
	Carbon calculators used to test GHG emissions show large differences, mainly due to how emissions from fertiliser manufacture and application are accounted for and whether LUC is excluded or incompletely calculated (Whittaker et al., 2013).	Although, cellulosic ethanol requires less fossil fuels for process heat and electricity than starch-based ethanol (AEA/NNFCC, 2010).	
		Biofuel producers may select a carbon calculator that generates the greatest GHG savings (differences in emission factors yield different results) to comply with sustainability criteria (Whittaker et al., 2013).	
Environmental impacts	Biodiversity and water preservation are seen as 'grand challenges' with far-reaching social ramifications.	Perennial energy crops can improve biodiversity and water quality due to the reduced requirement for nitrogen fertiliser and pesticide inputs; but slow-growing crops may affect groundwater recharge and require constant access to water (Karp et al., 2009).	Whole system water usage needs to be investigated (ideally across agriculture as a whole).

TABLE 1: *Cont.*

Sustainability challenge	First generation (1G)	Second generation (2G)	Policy challenges
	Concerns about the impacts of monocultures for biofuels on biodiversity and water conservation, especially linked to the conversion of natural terrestrial ecosystems, are not new.	Biodiversity and water impacts remain a concern: high-yielding food crops grown for their co-products and residues will use disproportionately more water while marginal or degraded land (Sims et al., 2010) or land formerly under the EU set-aside scheme (RCEP, 2000) may be targeted for monoculture energy crops.	Research on energy crop breed varieties that protect ecosystem services is ongoing, but differences between performance in laboratory conditions and ecological conditions in situ will need to be considered.
		Whole chain water use a concern (especially where 2G processing techniques are particularly water intensive).	Efforts to balance ecosystem services to preserve biodiversity are not new in UK agricultural systems (science stakeholder interview, 15 July 2011), however LUC impacts in the UK and abroad will add to policy complexity (science stakeholder interview, 5 October 2011).
			Impacts tend to be location-specific, so distribution of risks is an issue: yet impact assessments on biodiversity and water are constrained by considerable uncertainties including geospatial differentiation, different types of water and evapotranspiration (Jefferies et al., 2012).

TABLE 1: *Cont.*

Sustainability challenge	First generation (1G)	Second generation (2G)	Policy challenges
Other local impacts	Intensification of energy crops has been linked to long-term loss of livelihoods and local food/ energy production through displacement of local subsistence farmers (van Eijck and Romijn, 2008) and negative ecological impacts.	The visual impact of biomass (e.g., tall-growing miscanthus) and biomass plants on the landscape may be a factor, depending on the location.	Siting decisions for biofuel production facilities need to consider these local impacts, preferably in (early) consultation with the local community.
	Biomass plants can have a negative impact on local air quality through processing and transport emissions and on the aesthetics of the local landscape.	Processing and transport emissions remain a concern.	Understanding how public perceptions are shaped by broader social, cultural and personal meanings and assessments of bioenergy developments can help to develop more robust policy decisions, and social science can help in this regard.

Our map of sustainability issues arising from biofuels relies on the qualitative social research method of documents as a source of data and analysis (Bryman, 2012). We conducted a survey of articles in the field of energy research since the late 1970s, focusing on this flagship journal, supplemented by other key academic articles and reports produced by policy, professional and non-governmental organisations and the media. Treating these documents as a historical record of how debates about the sustainability of biofuels have evolved over time, we distilled the main themes, gaps or limitations and cross-cutting issues arising specifically around 1G biofuels. By comparison, there is less attention paid to 2G biofuel challenges in the documentary record, but we drew out the main themes where 2G was discussed.

We then tested and elaborated this map of challenges through semi-structured, in-depth interviews with 45 stakeholders from across the UK bioenergy 'system' (comprising science, industry, government and civil society, whilst recognising that some stakeholders may span more than one of these domains) to explore the state-of-the-art and future development of liquid transport biofuels in a global bioenergy system; and from a 2012 UK workshop involving 20 stakeholders that examined uncertainties inherent in life cycle assessment (LCA) of bioenergy and in estimations of the role of bioenergy in modelling the future UK energy mix (henceforth referenced as 'Modelling Uncertainties Workshop'). For the interviews, the established qualitative research approach of purposive sampling was used to sample stakeholders in a strategic yet sequential way, whereby an initial sample of stakeholders was selected by virtue of their relevance to the research questions posed, and the sample gradually added to as the investigation evolved (Bryman, 2012). This allowed a variety of stakeholder assessments from across the spectrum of the UK bioenergy system to be captured.

While our analysis focuses mainly on the UK context, since national and EU biofuel targets rely, implicitly or explicitly, on imports of biomass or biofuel rather than domestic supply, we refer to global issues where appropriate. Accordingly, the key challenges for policy that we pose are UK-focused, but may have broader relevance.

Our analysis draws on work in Science and Technology Studies (STS) (Rip, 1986, Cambrosio and Limoges, 1991 and Romijn and Caniëls, 2011) that argues that controversies fulfil an important technology assessment function in that they help articulate potential issues and problems that need to be considered in implementing new technologies. Irrespective of the validity of specific claims, controversies focus attention on key, often value-related, questions that were previously unrecognised and that need to be posed to address broader societal concerns. In line with Romijn and Caniëls (2011) who consider contestation and conflict as constitutive rather than constrictive of innovation systems, we suggest that controversies help to open up and expose the different elements of the socio-technical system or network which constitute a specific technology. Thus the controversy surrounding the development of particularly 1G biofuels has focused attention on the critical relationship between biofuels and sustainability that is shaping the limits of social acceptability of 2G biofuels.

The need for biofuel sustainability assessments to take into account the 'whole system' in an integrated manner is now generally recognised in numerous articles published in this journal and others such as Energy, and Renewable and Sustainable Energy Reviews. However, only a few of these focus specifically on lignocellulosic options for biofuels (e.g., Black et al., 2011, Haughton et al., 2009 and Singh et al., 2010). The state-of-the-art of whole system assessment of biofuels is also limited in a number of significant ways.

First, the social dimension is weakly integrated (if it is considered at all) into sustainability assessments which typically focus on LCA. Yet, from an overarching whole system perspective, there is a need to put these technical assessments in the broader context of social judgments that shape views on what is considered important and why. While some key publications do consider the social dimension, they also leave some gaps. Thornley et al. (2009) focus on constraints on UK biomass supply for bioenergy, whereas a whole system analysis needs to consider the role of imports in UK bioenergy policy and sustainability issues related to biomass conversion. The sustainability framework of Elghali et al. (2007) aims to take account of different stakeholder judgments but, as they observe, the method of ranking and weighing these through multi-criteria decision analysis (MCDA) is contested. Haughton et al. (2009) incorporate stakeholder views in their sustainability assessment framework; however, theirs is a case study of the biodiversity impacts of perennial crops in two specific regions in the UK while our assessment aims to examine a range of sustainability challenges for 2G biofuels (as a whole system from field to fuel) by drawing attention to the interface between the social dimension and the mainly environmental challenges of 1G and the potential implications for 2G.

Second, most sustainability assessments used in government policy (e.g., the 2012 UK Bioenergy Strategy) and in wider debate around biofuels focus on biomass supply to the relative exclusion of issues arising from the rest of the bioenergy chain (biomass pre-treatment and conversion through to bioenergy distribution and end-use). Consequently, although issues such as energy balance across the chain are usually considered in LCA, they are not widely discussed. In this respect, the whole system of bioenergy is not really considered, nor is the wider context of the social and policy system in which bioenergy research and policy are done. Our paper fills a gap in terms of bringing the sustainability of the bioenergy whole chain to bear on social judgments around biofuels.

Opening up the black-box of controversy surrounding 1G biofuels enables us to highlight a range of emerging challenges—encompassing the social, economic, ethical, ecological and political—that threaten to compromise perceptions of sustainability of 2G biofuels. The following section draws out and critically examines the key lessons that can be drawn from the controversy surrounding 1G biofuels, assesses their relevance for 2G, and highlights the key policy challenges in managing the transition to a sustainable UK bioenergy system. The key lessons arise from the most prominent themes that emerged from the documentary and stakeholder data and focus attention on the underexplored social dimensions in these areas. Thus we do not aim to map all the relevant sustainability challenges for biofuels—this has already been attempted by other authors (e.g., Markevičius et al., 2010 and Thornley and Gilbert, 2013).

11.3 THE RELEVANCE FOR 2G BIOFUELS OF SUSTAINABILITY CHALLENGES ARISING FROM 1G

Our analysis suggests that sustainability issues identified in relation to 1G are potentially relevant to 2G, and may become more prominent should

2G technologies be commercialised. In part this is due to some blurring of food and non-food biomass in current and future practice which is viewed positively by some (co-products of 1G crops and fuels can be 2G bioenergy feedstocks or used for animal feed) and negatively by others (intensification of agriculture implicates the production of residues as well as the crop). Our analysis suggests the need for a more comprehensive and integrated sustainability appraisal as the challenges are more complex than implied by the ubiquitous reference to 'food vs. fuel' conflicts. The implications of the retrospective analysis of 1G biofuels for the sustainability appraisal of 2G biofuels, including the challenges for policy in managing the transition, are summed up in Table 1 and further elaborated in the discussion below.

11.3.1 FOOD SECURITY

As early as 1991, Hall (1991: 733) noted that the food vs. fuel issue is 'far more complex than has been presented in the past and one which needs careful examination, since agricultural and export policies and the politicisation of food availability are greater determining factors'. However, the problem received scant policy attention until the 2007–08 world food price crisis that prompted warnings of sustained high food prices over the next decade as food production and supplies are displaced by biofuel production, particularly affecting developing countries that are net food importers (OECD/FAO, 2007). More recently, the negative impact of 1G biofuels on food security has been disputed (Pilgrim and Harvey, 2010) and the Nuffield Council on Bioethics (2011) notes that for every report or statement of a causal link between the 2007-08 spikes and biofuels, others provide rebuttals. However, the existence of multiple pressures on food prices such as rising meat consumption in the developing world may not exonerate biofuels. Searchinger (2011) argues 'if it is hard to meet rising food demands, it must be harder to meet demands for both food and biofuels'.

In the 2G case, depending on the type of land used and the suitability of this land for food crops, lignocellulosic energy crops potentially constitute a conflict with food production. In theory, bioethanol from 2G agricultural residues such as wheat straw should be exempt from such a conflict. In this standard account, there is a clear distinction between using land for human food consumption versus energy production. Bringing animal feed into the equation however reveals a more complex story. On the one hand, straw may be part of the animal feed mix thus representing a link to the food chain. This link may be an important consideration for barley and oat straw which, in the UK, are added as a source of roughage in livestock feed, but wheat-straw has less feed-value in this regard (Copeland and Turley, 2008). Practices may, however, differ in other agricultural systems in other countries and hence, this issue should be checked before it is certified free of conflicts with food. Some scientists and stakeholders setting out a case for lower meat consumption argue that reserving wheat straw (and land) for biofuels is still a better use than for animal feed, a key underlying value conflict in this debate that is beginning to emerge (Centre for Alternative Technology, 2010 and Carbon Cycles and Sinks Network, 2011).

11.3.1.1 POLICY CHALLENGES

'Food vs. fuel' could be a distorting simplification of the sustainability challenges raised by biofuels and one which overlooks the intrinsic interdependency of food and fuel; ergo, fuel is needed to produce food (Karp and Richter, 2011). Agricultural production of energy can complement food production by preventing or ameliorating rises in fuel and fertiliser prices that affect the food sector, suggest Murphy et al. (2011). But the efficacy of re-using marketable by-products and residues of bioethanol production relative to other methods for improving soil fertility should be considered (Singh et al., 2010). While acknowledging that food and fuel imperatives can conflict, Murphy et al. (2011) argue that we need better land management policies in order to reconcile and promote synergies between different uses of land for food and fuel.

The food–fuel conflict and the 1G/2G boundary are further complicated by conflicting value judgments over existing agricultural and land management practices as a whole. Using distillers' residues from ethanol production for animal feed may be credited as good waste management in appraising the use of a food crop for 1G fuel; however, this may be judged against the sustainability of the animal food and feed industry with some arguing it may be better to use land currently used for grain fed to animals, or indeed for grazing animals, for biofuel instead (Centre for Alternative Technology, 2010). The use of land for grazing or for grain fed to animals as opposed to direct human consumption could also be independently assessed (Wassenaar and Kay, 2008). Thus the use of land for fuel needs to be considered within a broader assessment of land use for different purposes, and how land is valued (Gamborg et al., 2012).

In addition to assessments of existing agricultural management practices, policy-makers also need to consider broader (industrial, residential and recreational, etc.) land use and management practices as a whole. Yet, debates on the 'sustainability of agriculture' and 'sustainability of land use' are lacking, even within the bioenergy community. While the possibility of such a broader debate is fraught by competing social, economic, environmental and policy arguments (and the implicit as well as explicit values embedded therein) it might entail, we would suggest that, given these complexities, there is a need for such a debate.

11.3.2 LARGE-SCALE LAND ACQUISITION

Linked to sustaining domestic food and energy security is the issue of land acquisition, especially on a large scale, both across and within national borders—and the uneven sustainability impacts this generates. Land in the global South acquired to produce biomass for fuel used in the global North or in domestic urban populations has been attracting critical scrutiny. The impacts of energy crop farming in developing countries have been argued to be both beneficial and harmful. Energy crops may provide income for the rural poor and lessen domestic dependency on fossil fuel imports while increasing opportunities for export revenue. But Doornbosch and Steenblik (2007) have questioned the ecological credentials of biofuels, asking 'Is the cure worse than the disease?' and highlighting local environmental harms to soil, water and biodiversity. NGOs have drawn attention to 'land grabs' leading to dispossession of local people and loss of livelihoods (Friends of the Earth, 2007 and Oxfam, 2007).

Land officially designated as marginal or degraded, but suitable for 2G feedstocks, may still be relied on to fulfil the livelihood, food and fuel needs of the rural poor. For agricultural residues, the sustainable use of land in which the crop is grown may be relevant, especially as such products are increasingly seen as global commodities and traded across national borders. Biofuel companies may also target land that promises higher returns on their investments such as agricultural or irrigated land where yields are higher (science stakeholder interview, 26 October 2011; civil society stakeholder interview, 20 December 2011), or forested land where additional revenues can be gained from logging (van der Horst and Vermeylen, 2011). The intensification of agriculture raises widespread concerns of negative ecological impacts including acidification, eutrophication, ecotoxicity and ozone depletion linked to deforestation and habitat loss (Doornbosch and Steenblik, 2007, Quadrelli and Peterson, 2007 and Tomei and Upham, 2009).

11.3.2.1 POLICY CHALLENGES

Given these problems, a key challenge for policy is to develop a framework for governing the practice of land acquisition in the global South. But in countries where social and environmental governance is weak, reliable monitoring of land acquisition is likely to be difficult given major power disparities between large companies and local people (van der Horst and Vermeylen, 2011). Relative priorities for different uses of land (food, fuel, grazing, recreation, biodiversity, etc.) are also likely to differ regionally or globally according to different communal or cultural value sets and need to be considered in all parts of the world (Thornley et al., 2009). Future land use priorities may also be shaped by the uneven distribution of global climate change impacts resulting in more or less emphasis placed on food or fuel production depending on local growing conditions.

11.3.3 GHG BALANCE

Depending on where they are grown, the land management practices, modes of transportation and processing techniques used, balancing the life cycle GHG emissions remains a challenge for 2G biofuels. In the case of straw, its removal is generally associated with negative impacts on soil nutrients and structure. While returning ash to the soil after combustion could compensate for these, the lost organic matter would affect the soil organic carbon content (Thornley et al., 2009). The processing of co-products such as wheat straw also poses significant sustainability challenges in that considerable energy is required to overcome the recalcitrance of lignocellulosic biomass through pre-treatment for enzymatic saccharification (Zhu and Pan, 2009). Techniques being explored to convert agricultural co-products, woody biomass or perennial crops are therefore also attracting increasing scrutiny. The conversion inefficiency of biomass for liquid biofuels has been raised by Clift and Mulugetta (2007) who make the point that higher efficiency and better GHG savings are possible for bio-heat or combined heat and power since biomass can be directly burned rather than converted to a liquid, a step that requires further energy inputs.

If 2G biomass were sourced by felling of forests, or where the use of marginal land which is in fact a source of food stimulates indirect land use change (iLUC) effects, then the GHG balance may be questioned. A detailed analysis of the effect of iLUC by Havlík et al. (2010), paying particular attention to the issues of deforestation, irrigation water use, and crop price increases due to expanding biofuel acreage, found that 2G biofuel production powered by sustainably sourced wood (rather than fossil fuels) would reduce overall emissions but that biomass feedstocks and land use may affect other sustainability criteria like biodiversity conservation, erosion protection, or even fuelwood supply for local communities.

Yet tools used to calculate GHG emissions vary according to scope, system boundaries and data sets that can lead to large differences in the results, as shown by Whittaker et al. (2013) in a study of UK feed wheat where GHG emissions from fertiliser manufacture and application are accounted for differently and where land use change (LUC) is excluded or incompletely calculated. To conform to sustainability criteria set by regulatory frameworks such as the EU's Renewable Energy Directive (RED), biofuel producers may select a tool that generates the greatest GHG savings.

11.3.3.1 POLICY CHALLENGES

Whether to include iLUC in GHG balance calculations is controversial. Given the complexity and uncertainty surrounding iLUC and its relevance to agriculture as a whole, deforestation (and the conversion of other forms of natural terrestrial ecosystems) may be more reliably addressed through dedicated policies that remove perverse incentives for biofuel production and reduce deforestation (wherever it occurs) through the development of strategies for sustaining forests and protecting biodiversity, rather than inclusion in GHG calculations for biofuels (Zilberman and Hochman, 2010). Assessing the energy balance of 2G biofuel production is also a central part of environmental appraisal. Here, there is some value in making the tacit assumptions and system boundaries underlying these calculations more explicit and reflecting on policy inferences from particular studies that may be more or less valid (Singh et al., 2010). This process may further blur the assumed distinction between 1G and 2G biofuels, but provides a more reasoned basis for preferring particular energy options for transport over others.

11.3.4 ENVIRONMENTAL IMPACTS

Biodiversity and water preservation are seen as 'grand challenges' that are likely to be intensified by the increasing demand for biofuels for transportation, with far-reaching social ramifications often at the local and regional scale. As early as 1991, David O Hall noted that monocultures for biofuels need to be reduced or avoided in order to maintain watersheds and ensure biodiversity. Similar concerns about monoculture plantations, especially in the global South, are echoed in recent critiques by NGOs (e.g., Action Aid, 2010).

Water is vital for maximising agricultural yield of crops grown for biomass and their residues. A study of the land and water implications of global 1G biofuel production in 2030 concluded that where traditional agricultural production already faces severe water limitations (such as in India and China, two of the world's largest agricultural producers and consumers), strain on water resources at the local and regional level would be substantial and policy-makers would be hesitant to pursue biofuel options based on traditional food and oil crops (De Fraiture et al., 2008). Where traditional food crops such as wheat need to be grown intensively if they are to meet the various demands of food, fuel, feed and fibre, Sinclair et al. (2004) have noted that yield and water are closely correlated, thus high yielding varieties will use proportionately more water. For this reason, agricultural wastes and residues are not immune from concerns about intensive water consumption.

To avoid land conflicts with food crops, marginal and degraded land could in theory be used for dedicated energy crops to be converted into 2G biofuels, however maintaining high yields over time is dependent upon continuous access to adequate water resources (Sims et al., 2010). Arable land formerly under the EU set-aside scheme to help preserve agricultural ecosystems may, if planted with monoculture energy crops, suffer a reduction in biodiversity (RCEP, 2000). Perennial energy crops can improve biodiversity and water quality due to the reduced requirement for nitrogen fertiliser and pesticide inputs but there is serious concern about the amounts of water needed by slower growing energy crops and the possible implications for stream flow and groundwater recharge (Karp et al., 2009).

The water intensity of biomass conversion, particularly lignocellulosic conversion to biofuel, is less well known than that of biomass cultivation. However, it has been suggested that methods of recycling water for re-use in processing systems, such as for evaporative cooling, are becoming increasingly sophisticated in modern ethanol plants (IEA, 2010). Lignocellulosic conversion can also produce waste streams that can be potentially harmful to water quality and the environment (science stakeholder interview, 2 August 2011). Concerns over the direct and indirect impacts of 2G biofuels on biodiversity and water conservation therefore warrant further investigation.

11.3.4.1 POLICY CHALLENGES

While attention has been paid to methods of recycling water for re-use in biofuel processing, whole system water usage needs to be investigated (ideally across agricultural systems as a whole). Research on energy crop breed varieties that do not compromise ecosystem services is ongoing in the UK, but differences between performance in laboratory conditions and ecological conditions in situ will need to be considered, especially as impacts will be felt most keenly at the local level. Efforts to balance ecosystem services to preserve biodiversity are not new in UK agricultural systems (science stakeholder interview, 15 July 2011), however LUC impacts linked to dedicated energy crops grown in the UK and abroad will add to policy complexity (science stakeholder interview, 5 October 2011). Biodiversity and water impacts, while of global concern, tend to be location-specific, so the distribution of risks brought about by iLUC, for example, needs to be considered in policy-making in the context of uncertainties that constrain impact assessments, including the difficulty in differentiating between geospatial regions, recognising different types of water and measuring evapotranspiration (Jefferies et al., 2012).

11.3.5 OTHER LOCAL IMPACTS

With the exception of GHG balance which is a global challenge unaffected by where emissions are produced or saved (Thornley and Gilbert, 2013), all the other impacts discussed so far can be described as 'grand challenges' whose impacts are experienced at a local level but where far-reaching, indirect social ramifications may also be felt. Human geographers highlight the necessarily spatially uneven character of sustainable transitions: that is, disproportionate (social, economic, environmental) burdens are placed on some social groups, places or ecologies, while sustainability elsewhere might be enhanced (Swyngedouw, 2007). Many of the direct impacts of biofuel production are likewise experienced at the local level especially where biomass is cultivated in poorer Southern regions for biofuel use elsewhere (van der Horst and Vermeylen, 2011).

There are widespread concerns that intensification, whether of 1G or 2G energy crops, will have negative ecological impacts including deforestation, habitat loss and declining soil fertility which in turn affect rural livelihoods. In their study of a developing biofuels sector based on Jatropha in Tanzania, van Eijck and Romijn (2008) recommend the development of policies to enhance the participation and capabilities of local communities in rural energy projects to ensure that sufficient attention is paid to their needs and preferences. Their study presents a cautionary tale for 2G that warns against a biofuels sector dominated by big commercial players interested in consolidating smaller holdings into larger plantations that will direct energy and financial revenues away from local communities. Any short-term profits offered to local farmers to sell their land to large investors will be seen as inadequate compensation for the long-term loss of livelihoods and local food and energy production.

Impacts on local communities are also a concern in the UK. Research on public attitudes to bioenergy in the UK has reported public concern about emissions and odours from bioenergy plants as well as the aesthetic impact on local landscapes (Barker and Riddington, 2003). Increased employment and financial returns to local farmers growing the biomass feedstock were seen as particular benefits of locally-sited bioenergy plants, although some concerns were expressed about the impact of heavy transport in the local area. These concerns remain relevant for 2G where, for example, the visual impact of biomass such as tall-growing miscanthus on the landscape may also be a factor depending on the location (Haughton et al., 2009). Although the higher density of woody biomass significantly reduces the need for transportation, thus limiting harmful emissions.

11.3.5.1 POLICY CHALLENGES

Siting decisions for biofuel production facilities need to consider these local impacts. Social science research has shown that public resistance to biomass development 'in their area' (Barker and Riddington, 2003) may be encountered, in particular where the public has not been properly consulted about the siting of a renewable energy development (Upham, 2009). A study of bioenergy developments in the Yorkshire and Humber region by Upham et al. (2007) concludes that pro-active exploration of public/ stakeholder attitudes and involvement may contribute to more strategic renewable energy planning at the local level. Yet, Devine-Wright (2008) notes that social scientific scrutiny into public opinion of renewable energy technologies, local resistance or acceptance, and the ways in which public engagement with these technologies is constructed and practised in the UK, is currently limited. Thus social science has a role in highlighting that renewable energy developments are situated in places—not 'sites'—that involve personal, local and cultural meanings and emotions as well as physical and material properties.

11.4 GAPS AND LIMITATIONS IN EXISTING STAKEHOLDER APPRAISALS THAT HAVE POTENTIAL IMPLICATIONS FOR 2G BIOFUELS

Drawing on stakeholder assessments identified in both the documentary review and stakeholder interviews, we call attention to a number of salient gaps and limitations in these appraisals of 1G biofuels that have potentially significant implications for the sustainability appraisal of 2G biofuels. Genetic modification and antimicrobial resistance are potentially important issues that have been neglected or somewhat marginalised in debates around biofuels, and may need to be considered. While the emerging impacts around these specific issues are difficult to quantify, they signify the importance of putting the appraisal of specific technologies in the broader context of alternative technology choices.

11.4.1 GENETIC MODIFICATION

Genetic modification (GM) techniques are seen by some as vital to achieving higher yields thereby increasing the net energy balance from energy crops through, for example, boosting resistance to pesticides and herbicides and to drought. The environmental impacts (both perceived and documented) of genetically modified organisms mean that GM techniques, even if they promise increased yields and a reduction in the need for crop protection, remain controversial in Europe (Gross, 2011). Yet controversy around GM may be the result of the socio-economic and political choices made rather than the technology in isolation (Levidow and Carr, 2009). However, among UK policy-makers there is currently little appetite to debate the use of GM in biofuel production while biofuels remain embroiled in debates over their broader sustainability credentials (policy stakeholder interview, 6 December 2011). The role of GM techniques in the reconstruction of 2G biofuels has received little attention (for an exception, see Levidow and Paul, 2008) by comparison with the role that synthetic biology might play in future generations, an option that has been covered widely in the media. This may yet become more widely debated if some 2G biofuels relying on GM for the development of energy crops and advanced processing techniques with improved yields become a reality. There is also the potential for going beyond the entrenched pro/anti GM debate by examining alternative uses of advanced genetics such as 'marker-assisted' plant breeding techniques (Stirling, 2013).

11.4.2 ANTIMICROBIAL RESISTANCE

Risk of antimicrobial resistance from the use of antibiotics in fermentation of ethanol has recently been highlighted within the scientific community (Muthaiyan et al., 2011). Where antibiotics are routinely used to control contaminants in bioethanol plants, antibiotic resistant bacteria may limit the effectiveness of antibiotics to treat future bacterial contamination. The by-products of grain processing for ethanol production, known as DDGS, contribute substantially to the economic viability of ethanol manufacturing. DDGS is increasingly used as an animal feed substitute for whole corn and soy and its potential application in other industries such as bioplastics renders it key to the long-term economic viability of the bioethanol industry. Since 2005 the EU has legislated against the use of animal feed products containing antibiotics residues as these can be directly harmful to cattle. Indirectly there may be a risk to human health if antibiotics enter the food chain through the consumption of crops fertilised with antibiotic laden manure or through absorption of antibiotics-contaminated water discharge from ethanol plants. There are concerns that the overuse of antibiotic agents in non-human settings in turn reduces the efficacy of antibiotics important for human medicine, as the antibiotics used are identical or nearly so (IATP, 2009).

11.5 CROSS-CUTTING CHALLENGES FOR THE WHOLE SYSTEM SUSTAINABILITY APPRAISAL OF 2G BIOFUELS

Table 1 focuses on specific, commonly understood sustainability challenges but highlights the complex interconnectedness of the economic, environmental and social dimensions of these. Our analysis suggests that there are also important cross-cutting issues that emerge between the lines, or are sometimes stated explicitly in documents or interviews but that have tended to be ignored in a focus on sustainability 'issues', which we now summarise. Although beyond the scope of the present study, the opportunities and challenges for the whole system sustainability appraisal of 2G biofuels represented by these cross-cutting issues require further investigation in their own right.

11.5.1 SPACE AND SCALE

A significant difference in the underlying framing assumptions of different stakeholder communities relates to the scale of biomass cultivation/sourcing and biofuel production. Civil society NGO stakeholders (interview, 20 December 2011; interview 20 February 2012) tend to be critical of large-scale (1G) biofuel developments due to concerns about intensification, monocultures, and human rights violations and difficulties of reliably monitoring complex North/South global supply chains and their local (spatial) impacts on sustainability (c.f. van der Horst and Vermeylen, 2011). Given that several policy scenarios (e.g., Bioenergy Strategy 2012; DECC Carbon Plan 2011) assume a key role for feedstock imports, including in the case of 2G, this is likely to remain a key issue. To highlight these problems, NGOs have tended to prefer the term 'agrofuels' rather than 'biofuels' (e.g., Action Aid, 2010). The environmental NGO, Journey to Forever (http://journeytoforever.org) argues that 'objections to biofuels as agrofuels are really just objections to industrialised agriculture itself, along with "free trade" (free of regulation) and all the other trappings of the global food system that help to make it so destructive'. By contrast, 'scaling up' to large industrial production units is seen by biofuel promoters to be essential for commercial reasons, but also one of the main challenges for 2G. In principle, a more inclusive dialogue on getting the logistics right (see Sanderson (2011) on concerns currently raised by biofuel experts on the need to locate biofuel production facilities close to the point of biomass cultivation) may help mediate some of this conflict, but this also needs to be supported by consideration of the social and political tensions raised by biomass as a global commodity. Thus policy and governance mechanisms for developing a bioenergy system at different scales and that are sensitive to different spatial impacts need to be examined.

11.5.2 TRADE AND ENERGY SECURITY

The initial case for 2G bioenergy rests on domestic (UK) feedstocks; in principle, this might be more stable, open to national control and less subject to international volatility once supply chains are established. In practice, biomass is an internationally traded commodity affected by a global market and WTO rules (policy stakeholder interview, 31 October 2011). Reliance on imported biomass is seen to be necessary in the short-term to meet UK and EU sustainable biofuels targets; however, even where there is a desire to change this practice, contractual terms may result in longer-term lock-in (industry stakeholder interview, 31 October 2011), an issue that would remain relevant for 2G feedstocks where these too are globally traded. Sustainability questions currently raised in connection to 1G are therefore likely to become relevant for 2G, for example, over competing uses of 'marginal' land for non-food energy crops or the use of residues by subsistence farmers (IEA, 2010). Where 2G feedstock trade is a more feasible option for some countries where there is no infrastructure for indigenous production, transferral of iLUC impacts to grower countries may occur. This raises the question of how trade in biomass versus trade in biofuel affects energy security concerns.

11.5.3 ENVIRONMENTAL IMPACT MODELLING

Recognition of the uncertainties, conditions and limits of the results from environmental modelling when making policy judgments and decisions on their basis was seen as crucial by stakeholders in the Modelling Uncertainties Workshop that we organised to explore issues around the interpretation of life cycle assessment (LCA) and bioenergy models. Stakeholders noted significant differences between attributional and consequential LCA1 in terms of levels of uncertainty, system boundaries and methodology. Yet the two approaches are sometimes conflated in the policy sphere (the EU's Renewable Energy Directive and the UK's Renewable Transport Fuel Obligation tend not to distinguish between the two), which can lead to the misuse of the data in policy analysis. Thus Stirling (2003) argues for the need to examine the assumptions made and boundaries drawn in quantitative environmental assessments. For example, the degree of potential reduction in GHG emissions by biofuels is dependent upon the biofuel feedstock, the management practices used and perhaps the nature (scale and distribution) of the industry (Schubert, 2006). Thus, rather than being intrinsic to the 'renewable' resource of biomass, environmental impacts depend also on a number of socio-technical practices. Scharlemann and Laurance (2008) use 'total' environmental impact assessments of different biofuel systems (considering potential losses of forests and farmland as well as GHG savings) to argue that some biofuels fare worse than fossil fuels. Stirling (2003) also argues that it is impossible to have a single standard of relative importance of impacts. The specific form of environmental impacts associated with different energy generating technologies, or even different food and fuel production processes, may be radically different. A key point that Stirling makes, which we echo, is the tendency to focus on impacts at the energy supply stage rather than in terms of subsequent use. For example, what will the relative impacts be of different transport energy choices on future air quality? Studies indicate that while combustion of renewable fuels may, in some cases, result in a reduction in regulated pollutants (e.g., CO_2), the emissions may contain significant amounts of currently unregulated yet equally important pollutants (Gaffney and Marley, 2009).

11.6 CONCLUSION

The controversy surrounding 1G biofuels has fulfilled an important technology assessment function (Rip, 1986, Cambrosio and Limoges, 1991 and Romijn and Caniëls, 2011) in that is has helped to articulate sustainability issues and challenges that need to be considered in implementing 2G biofuels. By drawing on the key lessons arising from 1G, we find that these are potentially relevant to the sustainability appraisal of 2G biofuels depending on the particular circumstances or conditions under which 2G is introduced. In doing so, we have highlighted the limitations of focusing on narrow framings or understandings of core sustainability challenges, such as the now ubiquitous 'food vs. fuel' conflict. Thus 'food vs. fuel' is a simplification of a complex array of interrelated factors not least to do with how land is valued, managed and governed.

A substantive lesson that we draw from opening up the different elements of the socio-technical system or network which constitutes (1G or 2G) biofuels, is acknowledging and understanding that challenges commonly categorised as the 'three pillars' of sustainability – economic, environmental, social – are in reality more complexly interconnected so that their artificial separation in sustainability appraisal is problematic. This point is vividly made in a study of the potential of growing perennial biomass crops, specifically SRC willow and miscanthus, for energy in the UK. Having described the potential benefits for energy security and climate policy, Karp et al. (2009) point out that these crops are physically different from currently grown arable crops – their harvesting patterns vary, they are very tall and dense, and have deeper roots, all of which has implications for a number of factors including the appearance of the rural landscape, tourist income, farm income, hydrology and biodiversity. The interrelationship of productive uses of land with the ecosystems, livelihoods and culture of specific locations, challenges notions that such connections can be simply erased and remade without cost or conflict and this has been evident in countries like the UK as well as in the global South.

At the beginning of this paper we argued the state-of-the-art of whole system assessment of biofuels was significantly limited by a tendency to focus on biomass supply to the relative exclusion of issues arising from the rest of the bioenergy chain, and by the weak integration, if at all, of the social dimension. The findings we have presented, culminating in the point made by Karp et al. (2009) above, demonstrate the importance for policy of considering the sustainability of the bioenergy whole chain in the broader context of social judgments around biofuels. To this end, we agree with Gibson (2006) who argues that an integrative understanding of sustainability appraisal calls for new forms of knowledge. Rather than treating sustainability as a matter of balancing or trading off different systems, such an approach would examine the interdependence of environmental, economic and social variables – the 'whole system'. While we cannot claim yet to have breached the disciplinary barriers, we have begun to lay the groundwork for a more integrated sustainability appraisal.

REFERENCES

1. AEA/NNFCC, 2010. Closing the loop: optimising food, feed, fuel and energy production oppportunities in the UK. http://www.nnfcc.co.uk/tools/closing-the-loop-optimising-food-feed-fuel-energy-production-opportunities-in-the-uk-spreadsheet-tool-nnfcc-10-015.
2. Action Aid, 2010. Meals per gallon, Action Aid, London. http://www.actionaid.org.uk/doc_lib/meals_per_gallon_final.pdf.
3. Barker, A.M., Riddington, C., 2003. Attitudes to Renewable Energy, MVA Project Number C32906, COI Communications, DTI, London. www.dti.gov.uk/renewables/renew_1.2.1.4a.htm.
4. M.J. Black, C. Whittaker, S.A. Hosseini, R. Diaz-Chavez, J. Woods, R.J. Murphy Life cycle assessment and sustainability methodologies for assessing industrial crops, processes and end products Industrial Crops and Products, 34 (2011), pp. 1332–1339
5. A. Bryman Social Research Methods (fourth ed.)Oxford University Press, Oxford (2012)
6. A. Cambrosio, C. Limoges Controversies as governing processes in technology assessment Technology Analysis and Strategic Management, 3 (4) (1991), pp. 377–396
7. Carbon Cycles and Sinks Network, 2011. The costs and benefits of moving out of beef and into biofuel. http://www.feasta.org/wp-content/uploads/2011/05/Biovs-beef_report.pdf.
8. Centre for Alternative Technology, 2010. Zero Carbon Britain 2030. http://www.zerocarbonbritain.org/.
9. M.B. Charles, R. Ryan, N. Ryan, R. Oloruntoba Public policy and biofuels: the way forward? Energy Policy, 35 (2007), pp. 5737–5746
10. Clift, R., Mulugetta, Y., 2007. A plea for common sense (and biomass). The Chemical Engineer. October, pp. 24–26.
11. Copeland, J., Turley, D., 2008. National and Regional Supply/Demand Balance for Agricultural Straw in Great Britain. http://www.nnfcc.co.uk/tools/national-and-regional-supply-demand-balance-for-agricultural-straw-in-great-britain.
12. C. De Fraiture, M. Giordano, Y. Liao Biofuels and implications for agricultural water use: blue impacts of green energy Water Policy, 10 (S1) (2008), pp. 67–81

13. P. Devine-Wright Reconsidering public acceptance of renewable energy technologies: a critical review M. Grubb, T. Jamasb, M.G. Pollitt (Eds.), Delivering a Low-Carbon Electricity System: Technologies, Economics and Policy, Cambridge University Press, Cambridge (2008), pp. 443–461

14. Doornbosch, R., Steenblik, R., 2007. Biofuels: is the cure worse than the disease? http://www.oecd.org/dataoecd/15/46/39348696.pdf.

15. L. Elghali, R. Clift, P. Sinclair, C. Panoutsou, A. Bauen Developing a sustainability framework for the assessment of bioenergy systems Energy Policy, 35 (2007), pp. 6075–6083

16. J. Fargione, J. Hill, D. Tilman, S. Polasky, P. Hawthorne Land clearing and the biofuel carbon debt Science, 319 (2008), pp. 1235–1238

17. Friends of the Earth, 2007. Biofuels—a big green con? http://www.foe.co.uk/news/biofuels.html.

18. J.S. Gaffney, N.A. Marley The impacts of combustion emissions on air quality and climate—from coal to biofuels and beyond Atmospheric Environment, 43 (2009) (2009), pp. 23–36

19. C. Gamborg, K. Millar, O. Shortall, P. Sandøe Bioenergy and land use: framing the ethical debate Journal of Agricultural and Environmental Ethics, 25 (2012), pp. 909–925

20. R.B. Gibson Beyond the pillars: sustainability assessment as a framework for effective integration of social, economic and ecological considerations in significant decision-making Journal of Environmental Assessment Policy and Management, 8 (3) (2006), pp. 259–280

21. M. Gross New directions in crop protection Current Biology, 21 (2011), pp. R641–R643

22. D.O. Hall Biomass energy Energy Policy, 19 (8) (1991), pp. 711–737

23. A.J. Haughton, A.J. Bond, A.A. Lovett, T. Dockerty, G. Sünnenberg, S.J. Clark, D.A. Bohan, R.B. Sage, M.D. Mallott, V.E. Mallott, M.D Cunningham, A.B. Riche, I.F. Shield, J.W. Finch, M.M. Turner, A. Karp A novel, integrated approach to assessing social, economic and environmental implications of changing rural land-use: a case study of perennial biomass crops Journal of Applied Ecology, 46 (2009), pp. 323–333

24. P. Havlík, U.A. Schneider, E. Schmid, H. Böttcher, S. Fritz, R. Skalský, K. Aoki, S. De Cara, G. Kindermann, F. Kraxner, S. Leduc, I. McCallum, A. Mosnier, T. Sauer, M. Obersteiner Global land-use implications of first and second generation biofuel targets Energy Policy, 39 (2010), pp. 5690–5702

25. House of Commons Environmental Audit Committee, 2008. Are biofuels sustainable? http://www.publications.parliament.uk/pa/cm200708/cmselect/cmenvaud/76/76.pdf.

26. IATP, 2009. Fueling Resistance? Antibiotics in Ethanol Production. Institute for Agriculture and Trade Policy. http://www.iatp.org/documents/fueling-resistance-antibiotics-in-ethanol-production.

27. IEA, 2010. Sustainable Production of Second-Generation Biofuels. Potential and perspectives in major economies and developing countries. http://www.iea.org/publications/freepublications/publication/biofuels_exec_summary.pdf.

28. D. Jefferies, I. Muñoz, J. Hodges, V.J. King, M. Aldaya, A.E. Ercin, L. Milà i Canals, A.Y. Hoekstra Water footprint and life cycle assessment as approaches to assess potential impacts of products on water consumption. Key learning points from pilot studies on tea and margarine Journal of Cleaner Production, 33 (2012) (2012), pp. 155–166

29. A. Karp, A.J. Haughton, D.A. Bohan, A.A. Lovett, A.J. Bond, T. Dockerty, G. Sünnenberg, J.W. Finch, R.B. Sage, K.J. Appleton, A.B. Riche, M.D. Mallott, V.E. Mallott, M.D. Cunningham, S.J. Clark, M.M. Turner Perennial energy crops: implications and potential M. Winter, M. Lobley (Eds.), What is Land For? The Food, Fuel and Climate Change Debate, Earthscan, London (2009), pp. 47–72

30. A. Karp, G.M. Richter Meeting the challenge of food and energy security Journal of Experimental Botany, 62 (2011), pp. 3263–3271

31. L. Levidow, S. Carr GM Food on Trial: Testing European Democracy Genomics and Society, Routledge, New York/London (2009)

32. L. Levidow, H. Paul Land-use, Bioenergy and Agro-biotechnology WBGU, Berlin (2008)

33. A. Markevičius, V. Katinas, E. Perednis, M. Tamašauskienė Trends and sustainability criteria of the production and use of liquid biofuels Renewable and Sustainable Energy Reviews, 14 (9) (2010), pp. 3226–3231

34. R. Murphy, J. Woods, M. Black, M. McManus Global developments in the competition for land from biofuels Food Policy, 36 (2011), pp. S52–S61

35. A. Muthaiyan, A. Limayem, S.C. Ricke Antimicrobial strategies for limiting bacterial contaminants in fuel bioethanol fermentations Progress in Energy and Combustion Science, 37 (2011), pp. 351–370

36. Nuffield Council on Bioethics, 2011. Biofuels: ethical issues. http://www.nuffield-bioethics.org/biofuels.

37. OECD/FAO OECD-FAO Agricultural Outlook 2007–2016 OECD Publications, Paris (2007)

38. Oxfam, 2007. Bio-fuelling poverty—why the EU renewable-fuel target may be disastrous for poor people. Oxfam Briefing Note. http://www.oxfam.org.nz/imgs/pdf/biofuels%20briefing%20note.pdf.

39. Pilgrim, S., Harvey, M., 2010. Battles over biofuels in Europe: NGOs and the politics of markets. Sociological Research Online, 15 http://www.socresonline.org.uk/15/3/4.html.

40. R. Quadrelli, S. Peterson The energy-climate challenge: recent trends in CO2 emissions from fuel combustion Energy Policy, 35 (11) (2007), pp. 5938–5952

41. Raman, S., Mohr, A.. Biofuels and the role of space in sustainable innovation journeys. Journal of Cleaner Production, http://dx.doi.org/10.1016/j.jclepro.2013.07.057, in press.

42. A. Rip Controversies as informal technology assessment Knowledge, 8 (2) (1986), pp. 349–371

43. H.A. Romijn, M.C.J. Caniëls The Jatropha biofuels sector inTanzania 2005–2009: evolution towards sustainability? Research Policy, 40 (4) (2011), pp. 618–636

44. RCEP, 2000. Energy—The changing climate. Royal Commission on Environmental Pollution, 22nd Report, London.

45. K. Sanderson Lignocellulose: a chewy problem Nature, 474 (2011), pp. S12–S14

46. Scharlemann, J.P.W., Laurance, W.F., 2008. Supporting Online Material for How Green Are Biofuels? www.sciencemag.org/cgi/content/full/319/5859/43/DC1.

47. C. Schubert Can biofuels finally take center stage? Nature Biotechnology, 24 (2006), pp. 771–784

48. T. Searchinger How Biofuels Contribute to the Food Crisis Washington Post (2011)

49. F. Sengers, R.P.J.M. Raven, A. Van Venrooij From riches to rags: biofuels, media discourses, and resistance to sustainable energy technologies Energy Policy, 38 (2010), pp. 5013–5027

50. R.E.H. Sims, W. Mabee, J.N. Saddler, M. Taylor An overview of second generation biofuel technologies Bioresource Technology, 101 (2010), pp. 1570–1580

51. T.R. Sinclair, L.C. Purcell, C.H. Sneller Crop transformation and the challenge to increase yield potential Trends in Plant Science, 9 (2) (2004), pp. 70–75

52. A. Singh, D. Pant, N.E. Korres, A.S. Nizami, S. Prasad, J.D. Murphy Key issues in life cycle assessment of ethanol production from lignocellulosic biomass: challenges and perspectives Bioresource Technology, 101 (2010), pp. 5003–5012

53. A. Stirling Renewables, sustainability and precaution: beyond environmental cost-benefit and risk analysis Issues in Environmental Science and Technology, 19 (2003), pp. 113–134

54. Stirling, A., 2013. Why all the fuss about GM food? Other innovations are available. Political Science Blog, The Guardian, http://www.theguardian.com/science/political-science/2013/jun/28/gm-food.

55. E. Swyngedouw Impossible 'sustainability' and the postpolitical condition R. Krueger, D. Gibbs (Eds.), The Sustainable Development Paradox: Urban Political Economy in the United State and Europe, Guildford Press, New York (2007), pp. 13–40

56. P. Thornley, P. Gilbert Biofuels: balancing risks and rewards Interface Focus, 3 (1) (2013), pp. 2042–8901

57. P. Thornley, P. Upham, J. Tomei Sustainability constraints on UK bioenergy development Energy Policy, 37 (2009), pp. 5623–5635

58. J. Tomei, P. Upham Argentinean soy-based biodiesel: an introduction to production and impacts Energy Policy, 37 (10) (2009), pp. 3890–3898

59. P. Upham Applying environmental-behaviour concepts to renewable energy siting controversy: reflections on a longitudinal bioenergy case study Energy Policy, 37 (2009), pp. 4273–4283

60. P. Upham, S. Shackley, H. Waterman Public and stakeholder perceptions of 2030 bioenergy scenarios for the Yorkshire and Humber region Energy Policy, 35 (9) (2007), pp. 4403–4412

61. D. van der Horst, S. Vermeylen Spatial scale and social impacts of biofuel production Biomass and Bioenergy, 35 (2011), pp. 2435–2443

62. J. van Eijck, H. Romijn Prospects for Jatropha biofuels in Tanzania: an analysis with strategic Niche management Energy Policy, 36 (1) (2008), pp. 311–325

63. T. Wassenaar, S. Kay Biofuels: one of many claims to resources Science, 321 (2008), p. 201

64. C. Whittaker, M. McManus, G.P. Hammond Greenhouse gas reporting for biofuels: a comparison between the RED, RTFO and PAS2050 methodologies Energy Policy, 39 (2011) (2011), pp. 5950–5960

65. C. Whittaker, M. McManus, P. Smith A comparison of carbon accounting tools for arable crops in the United Kingdom Environmental Modelling and Software, 46 (2013) (2013), pp. 228–239

66. J.Y. Zhu, X.J. Pan Woody biomass pretreatment for cellulosic ethanol production: technology and energy consumption evaluation Bioresource Technology, 101 (2009), pp. 4992–5002

67. D. Zilberman, G. Hochman Indirect land use change: a second-best solution to a first-class problem Fuel, 13 (2010), pp. 382–390

Author Notes

CHAPTER 2

Funding
This research was funded by the U.S. Department of Agriculture (USDA), Agricultural Research Service (ARS) including funds from the USDA-ARS GRACEnet effort, and partly by the USDA Natural Resources Conservation Service. The funders had no role in study design, data collection and analysis, decision to publish, or preparation of the manuscript.

Competing Interests
The authors have declared that no competing interests exist.

Acknowledgments
Authors would also like to acknowledge the contribution of Edward J. Wolfrum and his staff at the National Renewable Energy Laboratory (Golden, CO) in conducting the corn stover analysis. Mention of trade names or commercial products in this publication does not imply recommendation or endorsement by the U.S. Department of Agriculture. USDA is an equal opportunity provider and employer.

Author Contributions
Conceived and designed the experiments: KV GV RF. Performed the experiments: KV GV RF RM MS VJ. Analyzed the data: MS. Contributed reagents/materials/analysis tools: MS KV RM RF. Wrote the paper: MS KV GV RF RM VJ.

CHAPTER 4

Conflict of Interests
The authors declare that there is no conflict of interests regarding the publication of this paper.

Acknowledgments

Two anonymous referees and the editor of this paper are thanked for their suggestions for the improvement of the paper. The study was supported by the Chinese Academy of Sciences (Grant no. KZZD-EW-08) and the High Resolution Earth Observation Systems of National Science and Technology Major Projects (05-Y30B02-9001-13/15-10).

CHAPTER 5

Funding

This study was supported by the University of Minnesota and the Minnesota Environment and Natural Resources Trust Fund as recommended by the Legislative-Citizen Commission on Minnesota Resources. The funders had no role in study design, data collection and analysis, decision to publish, or preparation of the manuscript.

Competing Interests

The authors have declared that no competing interests exist.

Acknowledgments

We would like to thank K. Johnson, M. DonCarlos and A. Rasmussen for multiple years of diligent field work. We are grateful for the logistic support provided by the MNDNR Talcot Lake Wildlife Management Area. We would also like to thank two anonymous reviewers for thoughtful comments and faculty and students at the University of Minnesota Conservation Biology Graduate Program.

Author Contributions

Conceived and designed the experiments: JJ CS DW CL. Performed the experiments: JJ. Analyzed the data: JJ. Contributed reagents/materials/analysis tools: JJ JF CS DW CL. Wrote the paper: JJ.

CHAPTER 7:

Conflict of Interest

None declared.

CHAPTER 8

Acknowledgments
This work was supported by the United States Department of Energy, Office of the Biomass Program under Contract No. DE-AC36-99GO10337 with the National Renewable Energy Laboratory (NREL).

Conflict of Interest
The authors declare that they have no conflict of interest in this publication.

CHAPTER 9

Acknowledgements
The authors gratefully acknowledge financial support from the EKO-BENZ Ltd

CHAPTER 10

Acknowledgments
This study was supported by the Biomass Program in the US Department of Energy's Office of Energy Efficiency and Renewable Energy under Contract DE-AC02-06CH11357. We are grateful to Zia Haq and Kristen Johnson of the Biomass Program for their support and guidance. We thank the two reviewers of this journal for their helpful comments. The authors are solely responsible for the contents of this article.

CHAPTER 11

Acknowledgements
The research reported here was supported by the Biotechnology and Biological Sciences Research Council (BBSRC)Sustainable Bioenergy Centre (BSBEC), under the programme for 'Lignocellulosic Conversion to Ethanol' (LACE) [Grant Ref: BB/G01616X/1]. This is a large interdisciplinary programme and the views expressed in this paper are those of the

authors alone, and do not necessarily reflect the views of collaborators within LACE or BSBEC, or BBSRC policies. We acknowledge the contribution of Robert Smith who conducted the interviews with civil society stakeholders cited in the paper. The paper has also benefitted from the helpful suggestions of an anonymous reviewer.

Index

A

acetic acid, 176, 178, 212
aerosols, 192
aggregation, 139
air quality, 27, 287, 304, 307
alfalfa, 58
alkali metals, 102
allocation, 8–9, 12, 24, 63, 252,
  261–263, 269, 275, 279
amortization, 259
antibiotics, 301, 307
antimicrobial resistance, 300–301
arabinose, 37, 107, 176
arable, 9–10, 18, 33, 67, 73, 76, 85,
  297, 305, 310
ash, 102–103, 192, 202, 295
atmosphere, 178, 187, 190–191,
  201–202, 205–206, 221, 235,
  240–241, 258

B

beef, 58, 60, 67–68, 77, 79, 81–82, 306
biodiesel, xix, 3, 9, 12, 15, 17–26,
  28, 30–31, 51, 80, 84, 86, 91, 93,
  96–100, 128, 151–153, 155–157,
  159–161, 163–165, 167, 169–174,
  204–205, 227–229, 231–232, 235,
  246–247, 249, 309
biodiversity, xvii, 4, 11, 33, 47, 80,
  89, 101–102, 128, 130, 132, 228,
  240–241, 284–286, 289, 293–298, 305
bioenergy, xvi–xix, xxi, 3, 11, 26,
  28–31, 33–35, 42, 45–49, 83–91,
93, 95–99, 101–104, 107, 116–118,
122–123, 125–133, 135–137, 139,
141, 143–147, 228–231, 275–279,
282–283, 287–291, 293, 299,
302–305, 307–309, 313
  bioenergy potential, xviii, 84–85,
  97–98, 102–104, 107, 116–118,
  126–127, 132–133, 144
biofuel
  algae-derived, 151
  biofuel demand, 9, 19, 23, 25, 33
  biofuel development, 10, 14, 17,
  22, 97
  biofuel feedstock, 10–11, 27, 49,
  51, 53, 55, 57, 59, 61, 63, 65, 67,
  69, 71, 73, 75, 77–81, 86, 258,
  277, 304
  biofuel policy, 4, 17, 22, 29
  biofuel production, xvi, 11, 13–14,
  20, 22, 24, 26, 28–30, 40, 49, 76,
  78, 80, 97, 103, 128, 147, 206,
  222, 224, 226, 228, 230, 232, 234,
  236, 238, 240, 242, 244, 246, 249,
  251, 258–259, 279, 283, 287, 291,
  295–296, 298–299, 301–303, 309
  biofuel sustainability, 289
  biofuel volume, 259
  cellulosic biofuel, 34, 46, 278
  first generation biofuels (1G
  biofuels), xvi–xvii, xx–xxi, 3, 9,
  11, 19, 26, 47, 204, 281–283, 285,
  287–293, 295–305, 307, 309
  second-generation biofuels (2G
  biofuels), xv, xx–xxi, 49, 86, 98,

149, 245, 282–283, 286, 288–292, 294–303, 305, 307

biomass
  agricultural biomass, 199
  biomass cultivation, xviii, 10, 132, 297, 302–303
  biomass electricity, xix, 132
  biomass energy, 38, 96, 143, 147, 199, 202, 244, 307
  biomass harvest, 102, 111, 121, 144
  biomass productivity, 106, 129
  biomass quality, xviii, 102, 115, 123, 126
  biomass storage, 144
  biomass transport, 37, 144
  biomass yield, xviii–xix, 36, 40, 49, 103–104, 106–108, 110–127, 130, 133, 135, 146, 180, 273
  forest biomass, 199, 204
  harvested biomass, 37, 40, 102, 106, 110, 113, 117, 120–121, 232
  perennial biomass, xvii, 101, 127, 305, 307
  solid biomass, 178, 200–202
  waste biomass, 144, 146, 173, 206, 238
biorefinery, xvii, 36, 47, 120, 123, 170, 206, 210

C

Canadian Economic and Emissions Model for Agriculture (CEEMA), 58–59, 61–63, 65, 67, 72, 76, 78–79
Candida antarctica, 169
canola, 51, 58, 78, 229
carbohydrate, 37, 86, 103, 106–107, 201
carbon capture and storage (CCS), 206
carbon
  carbon dioxide (CO_2), xvi–xvii, xx, 8, 19, 21–25, 27, 40, 42, 45, 50–54, 57, 59, 70–80, 84, 146, 167, 187–189, 191, 193, 195–199, 201–203, 205–207, 209–219, 221–223, 225–229, 231–233, 235, 237, 239–248, 251, 262, 264, 266, 275, 278, 304, 308
  carbon emissions, 132, 258, 277
  carbon stocks, 10–11, 284
carotenoids, 196, 198
Cassava, xvii, 85–86, 90–92, 94–99
catalyst, 162–163, 165, 167–169, 173, 195, 213–216, 218, 220–224, 226–227, 239, 246–248
cell wall, 37, 40, 106–107, 176, 232
cellulose, 38, 43, 49, 103, 176, 186, 205, 209, 257
cellulosic, xvi, xix–xx, 34, 36–40, 42–43, 45–51, 128, 147, 209, 249–252, 257–258, 260, 263, 265, 267, 271–274, 276, 278, 282, 285, 310
census, 54, 56, 59, 62–63, 73, 77, 79, 82, 139
chemical oxygen demand (COD), 30, 155
chlorophyll, 195–197, 214, 228
climate, xvi, 3, 13, 23, 29, 31, 33, 39, 45, 47, 53, 69, 76, 79, 83, 89, 91, 99, 115, 126–127, 174, 243–245, 264, 271, 276–277, 284, 294, 305, 307–308
  climate change, xvi, 3, 13, 39, 45, 53, 83, 89, 243, 264, 276–277, 284, 294, 308
  climate change mitigation, 83, 276
co-products, 8–9, 17, 24, 38–39, 251, 262–263, 278, 282, 286, 291, 295
coal-fired plants, 70, 229
coefficient of variation (CV), 67–69, 106
combined heat and power (CHP), 36, 257–258, 273, 295
conservation, xvii–xviii, 45, 49, 101–105, 107, 109, 111, 113–115,

117–121, 123–129, 276, 286, 295, 297, 311–312
    Conservation Reserve Program (CRP), 45, 49, 102–104, 127–128
construction, 86, 195, 210, 232, 244
controversy, xx–xxi, 282, 289–290, 300, 304, 306, 308–309
cooking oil, 152, 171, 173
corn (*Zea mays* L.), xix–xx, 9, 11, 30, 33–51, 58, 63, 74, 81, 84–85, 101, 147, 175–179, 181–183, 185–186, 246, 249–252, 254–256, 258–260, 263, 265–267, 269–279, 301, 311
    corn ethanol, 48–50, 250–252, 254–256, 259, 263, 265, 267, 269–272, 275, 278–279
    corn starch, 49, 249
    corn stover, xix–xx, 34–37, 39–40, 42, 45, 47–49, 74, 175–179, 181–183, 185–186, 251–252, 256, 258–260, 263, 265–266, 269–271, 273, 276, 278, 311
    corn stover cell wall, 37
cost, 34, 36–37, 39, 47, 50, 52, 56, 77–78, 90, 96–97, 126, 128, 153, 156, 161, 167–168, 174, 202–203, 221, 235, 243, 245, 272, 277, 284, 305–306, 309
    cost-benefit analysis, 90
    production costs, 36
crops
    crop displacement, 12, 28
    crop management, 10, 18
    crop residues, 86, 250
cropland, xix, 11, 18–19, 22, 25, 31, 33–34, 41, 43–44, 46–47, 77, 90, 133–134, 137–140, 143–144, 147, 258
    marginal cropland, 34

D

delignification, 177–178, 181

development, xv, xvii, 3, 8–11, 14–18, 22, 26–27, 34, 51, 78, 80–81, 84–86, 88, 90, 96–97, 99, 129, 132, 146, 153, 187, 205–206, 210–211, 223, 231, 240, 242, 248–249, 275, 279, 281, 288–289, 296, 298–299, 301, 309
dimethyl sulfoxide (DMSO), xix–xx, 175–184
disaggregation, 54, 56, 59–61, 67–69, 71–72, 77–79
drought, 77, 85, 91, 93, 252, 273, 300
    drought resistance, 91
dry distillers grain (DDGS), 36, 38–39, 301

E

economic, xvi, xviii, xxi, 4, 10, 12, 15–16, 18–19, 21, 24, 27, 46–47, 49, 52, 58–59, 76, 81, 85, 91, 96, 98, 101–102, 128, 132, 187, 203, 211, 221, 249, 277–278, 282, 290, 293, 298, 300–302, 305–307
    economic development, 187, 249
    economic equilibrium, 4, 15, 27
    economic viability, xviii, 101, 301
ecosystem, 4, 9, 27, 33, 47, 80, 101, 125, 130, 241, 286, 296–298, 305
effluent, 37, 216
elasticity, 15–16, 26
electricity, xix, 36–39, 53, 56–57, 69–70, 131–132, 135–136, 141–143, 145, 147, 194, 200–202, 204, 207, 209, 224–225, 231, 251, 255, 257–258, 260, 263, 265, 267, 271–273, 285, 307
    electricity demand, 135–136, 141, 143
elevation, 88–89, 91
energy
    energy balance, 192, 202, 221–222, 244, 265–266, 284–285, 290, 296, 300

energy budget, xvii, 52, 54, 56–57, 59, 65–67, 69–72, 75–76, 78–79

energy conversion, 56, 125, 132, 135, 144, 201, 208

energy dependency, 86

energy dynamics, 70

energy intensity, 58, 67, 265, 273

energy markets, 84

energy ratio, 76, 265

energy ratios, 266

energy security, 3, 26, 83, 249, 293, 303, 305, 308

energy storage, xix, 131–133, 135–137, 139, 141, 143, 145–147

renewable energy, xviii, 19, 21, 28, 37, 46, 83–84, 96, 131, 135, 143, 146, 156, 175, 186, 188, 227, 245–246, 251, 276–277, 295, 299–300, 304, 306–307, 309, 311, 313

solar energy, 131–132, 135–136, 143, 146, 187–192, 194–196, 199, 201, 203, 210, 215, 235, 238, 242–244

environmental impacts, xvi, 4, 6, 8, 12, 27, 31, 80, 83, 98, 147, 285, 296, 300, 303–304

environmental performance, 4, 6, 9, 27

equilibrium models, 4, 15–16, 18–19, 27

general equilibrium models, 15–16

esterification, 152, 161–162, 164, 167, 170, 173–174, 177, 184

ethanol, xvi–xx, 3, 9, 19, 24–25, 29, 31, 34–52, 75, 84, 90, 96–99, 103–104, 107–108, 111–112, 114–119, 121, 123, 125–129, 146–147, 175–176, 178–179, 186, 203–204, 207, 210, 220, 222–223, 232, 242, 246–247, 249–261, 263–267, 269–279, 282, 285, 292, 297, 301, 307, 309–310, 313

cellulosic ethanol, xix, 36–40, 43, 45–46, 48–51, 147, 251–252, 257–258, 260, 263, 265, 267, 271–274, 276, 278, 282, 285, 310

corn-based, xx, 9, 47, 251, 254, 274

ethanol conversion efficiency, xviii, 40, 103–104, 107–108, 111–112, 115–119, 121, 123, 125–127

ethanol manufacturing, 51, 301

ethanol recovery, 37, 40

grain ethanol, xvi–xvii, 34, 36–37, 39–40, 42, 45–47, 51–52

eutrophication, 27, 229, 294

F

farm, xvii, 35–37, 39, 51–63, 65–82, 107, 202, 232, 242, 273, 305

farm energy budget, xvii, 52, 54, 56–57, 59, 65–67, 69–71, 75–76, 78–79

farm energy consumption, 51, 59, 63, 78

farm field operations, 53–54, 63, 80

farm field work, 53, 56, 59, 61, 66, 68–69, 71

farm machinery, 36, 39, 53, 56, 66–67, 70, 79–80, 202

farm types, 55, 57–63, 65–70, 73–75, 77–78

livestock farms, 57

fats, oils, and greases (FOG), xv, xix, 3, 152–157, 159–173, 205, 209

fatty acid methyl ester (FAME), 23, 152, 161–165, 167–169, 171, 173, 204, 232

feedstock production, xvi, 8–10, 27, 34, 49, 51–53, 72, 75, 77, 80, 127, 273, 277

fertilizer, xvi, 34–36, 39–40, 42, 44, 56, 58, 63, 67, 70–71, 77, 79, 82, 126, 202, 229, 234–235, 254–257, 260, 265, 267, 271, 273, 275, 284

nitrogen fertilizer (N fertilizer),
    34–36, 39–40, 44, 56, 58, 77, 126,
    254–255, 257, 260, 271, 273
filter, 167, 179, 256–257
    filter cake, 179, 256–257
flexible-fuel vehicles, 250
food
    food chain, 199, 283, 292, 301
    food economies, 132
    food security, xv, 84–85, 89, 97,
        282–283, 291
    food services, 156
    food vs. fuel, 282–283, 291–292, 305
forb cover, xviii, 116–118, 123–125
free fatty acids (FFA), 152, 158–159,
    161–164, 167–171, 173
frost, 34–35, 37, 40–43, 121
functional group, 107, 110, 116, 126,
    130

G

gasoline, xx, 9, 34, 39–40, 53, 56–57,
    59, 61, 68–70, 72, 78–79, 203, 210,
    213, 219–220, 222–223, 232, 250–251,
    254, 261–267, 269–272, 274, 277
genetic modification (GM), 129, 242,
    300–301, 308–309
genetics, xiv, 33, 301
Geographic Information System
    (GIS), xvii, 86, 89, 95, 97, 133
glacial till, 104
global positioning system (GPS),
    106–107
global warming potential (GWP), 8, 84
glucan, 37
glucose, 107, 195–196, 231, 238
glycerin, 152, 167
grease, xix, 151–157, 159–161,
    163–165, 167–174
    brown grease, xix, 152–153, 167,
        171, 173

grease abatement, 153–154, 159
grease traps, 153–157, 160, 171,
    173–174
yellow grease, 152, 159, 161
greenhouse gases (GHG)
    greenhouse gas emissions, xx, 3, 9,
        11, 13, 16–18, 24, 27–28, 30–31,
        33–37, 39, 41, 43–50, 52–53,
        58–59, 79–82, 84, 86, 101, 249,
        251–255, 257–259, 261, 263–269,
        271–279, 285, 295, 304
    GHG sequestration, 251, 259
    GHG sink, 39

H

Harmonized World Soil Database
    (HWSD), 89, 99
harvest, xvi, xviii, 34–35, 37, 40–45,
    48, 102–105, 110–111, 113, 116,
    120–121, 123–125, 127–130, 144,
    257, 271, 274, 278
    harvest intensity, 110–111
heating, xix, 38, 53, 56–57, 59–61, 66,
    68–72, 78–79, 180, 182, 184, 203,
    207, 209, 213, 239, 265
    heating fuels, 56–57, 59, 61, 68–72,
        78–79, 207
hemicellulose, 103, 176, 186, 257
herbicides, 36, 39, 105, 300
hydrodeoxygenation, 169, 174

I

impact analysis, 8, 31, 276
irrigation, 256, 295

J

Jatropha curcas L (JCL), xvii, 85–86,
    91, 93–96

K

kerosene, 214

L

land
   abandoned, xviii–xix, 85, 97,
      131–135, 137–140, 143–144, 147
   cropland, xviii–xix, 11, 18–19,
      22, 25, 31, 33–34, 41, 43–44,
      46–47, 49, 77, 90, 131, 133–135,
      137–141, 143–147, 258, 278
   land acquisition, 283, 293–294
   land availability, xix, 16, 99, 128,
      133–134, 138, 147
   land capabilities, 51
   land cover, xvii, 16, 18, 24, 86, 88,
      90, 94–96, 134, 147
   land degradation, 33
   land ethanol yield, xviii, 108,
      114–115, 123, 125–126
   land expansion, 12, 33
   management, xvi, 6, 10, 13, 18, 30,
      33–35, 40, 42, 45–46, 48, 52, 65,
      78–79, 82, 96, 102–104, 126–129,
      154, 283, 292–294, 304, 306–307,
      309, 312
   marginal land, xvii, 25, 83, 85–99,
      295
   pasture, 10, 58, 72, 121, 129,
      133–134, 138, 140, 143–144, 258
   rangeland, 63
   shrub land, xvii, 90, 94–96
   unused land, 88, 90
   wasteland, 90
land use, xv–xvi, xviii–xx, 1, 3–5, 7,
   9–11, 13–15, 17–19, 21, 23, 25–31,
   39, 46–47, 49–53, 58–59, 63, 65, 75,
   77–80, 101, 132–134, 138–139, 143–
   144, 146–147, 251, 258–259, 261,
   275–278, 283, 293–295, 307–308, 310

land use changes, xvi, 4, 9–10, 18,
   26–27, 29–31, 39, 75, 278
   land-use modeling, xix, 132, 139
landfill, 155, 205, 211, 245
laucustrine, 104
legume, 104, 110, 116–119, 121,
   123–127, 129–130
   legume cover, 116–117, 119,
      125–126
life cycle assessment (LCA), xvi, 4,
   6, 8–9, 14–16, 21, 23–24, 26–31, 36,
   46–47, 82, 84, 96–98, 146–147, 151,
   170, 206, 254, 259, 262–263, 272, 275,
   277–279, 288–290, 304, 306, 308–309
   attributional (aLCA), 4, 8–9,
      23–24, 304
   LCA methodology, 9, 24, 259
   prospective, 8, 17, 28, 206, 210
   cLCA, 8–9
lignin, 38, 42–43, 103, 176–177, 181,
   206, 251, 257–258, 263
lignocellulose ratios, 103
linear regression, 108
linoleic acid, 157–159
liquid propane (LPG), 57, 60, 66,
   70–71
litter, 107, 120
Livestock Crop Complex (LCC),
   58–59, 62–64

M

machinery, 36, 39, 53, 56, 66–68,
   70–71, 79–80, 202, 245
magnetic field, 168–169
manufacturing, 51, 53, 70, 78, 80, 202,
   216, 243, 301
media, 245, 281, 288, 301, 309
meta-analysis, 24
methane (CH$_4$), 53, 155, 201, 209,
   220, 226–227, 239–240, 247–248,
   254, 261–262, 264, 274

methane flaring, 261
methanol, 161–164, 167–170, 205, 210,
    212, 215–216, 218–223, 242, 246
milkweed (*Asclepias syriaca*), 124
Miscanthus, xx, 135, 145, 147, 251–252,
    256, 258–260, 263, 265–267, 269–270,
    272–273, 279, 282, 287, 299, 305
modeling, xvi, xix, 15, 18, 24, 27, 30,
    56–57, 78, 103–104, 119, 132–133,
    139, 147, 252, 254, 258–259, 261,
    266–267, 276–277
    stochastic modeling, 254
moisture, 103, 125, 168, 202, 256
monoculture, 101, 286, 296–297
multi-criteria decision analysis
    (MCDA), 289

N

natural gas (NG), 36–37, 42–43,
    47, 56–57, 60, 66, 71, 76–77, 81,
    121, 155, 173, 202, 205–206, 240,
    254–255, 263–265, 275
near infrared spectroscopy (NIRS),
    49, 106
net energy yield (NEY), 36, 38, 42–43,
    45, 47, 49
nitrogen (N), xvi, xviii, 23, 28, 34–36,
    39–45, 47–49, 56, 58, 62, 70, 77,
    80–82, 98–99, 103–104, 106–108,
    110–112, 115–121, 125–130, 147,
    160, 169, 171–172, 174, 190, 216,
    220, 226, 229, 231, 238, 241, 254–
    257, 260, 271, 273, 275, 284–285,
    297, 306–307, 309
nitrous oxide (N_2O), 8, 26–27, 29, 39,
    45, 53, 202, 254–257, 260, 264, 267,
    271–273

O

oilseed, 58–61, 63, 68, 77, 96, 229, 241

olefins, 170, 223
ozone (O_3), 190–191, 294

P

palm oil, 12, 23, 98
pasture, 10, 58, 72, 133–134, 138, 140,
    143–144, 258
pesticides, 56, 70, 160, 202, 285, 297,
    300
petroleum, xv–xvi, xx, 3, 33–36,
    38–39, 42, 44–45, 83, 151, 187, 254,
    261–267, 271, 274–275, 278–279
    petroleum offset, xvi, 34–36,
        38–39, 42, 44–45
pH, 28, 30, 89, 107, 110, 118, 124, 228
photosynthesis, 189, 194–196, 198–
    199, 201, 214–215, 218–219, 226,
    228, 232, 235, 238, 242–244, 246
plant composition, 103
plant tissue N, xviii, 103–104, 107,
    111–112, 125
policy, xvi, xviii, xxi, 3–4, 15, 17, 22,
    26, 29–31, 53, 81, 90, 95–96, 98,
    100, 104, 128, 132, 144, 170, 277–
    278, 281–283, 286–294, 296–309
poplar, 84–85
population, 59, 63, 76, 79, 84, 147,
    151, 188, 211
pork, 58, 67–68, 77, 79, 82
poultry, 58, 60, 67–68, 77, 82
precipitation, xviii, 88–89, 91, 94, 104,
    110–111, 115, 117, 121–122, 124,
    179
pressure, 11–12, 27, 85, 178, 203,
    211–212, 216, 220–222
price, 10, 15–16, 25, 83, 203, 250,
    283, 291, 295
    price changes, 15, 25
production, xvi, xviii–xx, 6, 8–16,
    18–20, 22, 24–31, 33–36, 38–40,
    45–53, 58–59, 63–64, 72, 74–85,

89, 93, 96–99, 102–104, 107, 121, 123–124, 127–133, 135–137, 139, 141–147, 155, 170–177, 187–189, 191, 193–195, 197, 199, 201–207, 209–211, 213–219, 221–247, 249–252, 254–262, 265, 267, 271–279, 283, 285, 287, 291–292, 294–296, 298–299, 301–304, 306–310
public policy, 26, 306

R

rapeseed, 12, 15, 18–19, 21–23, 204, 227
raw materials, 6, 17–18, 23, 30, 97, 203–204, 209, 213, 240, 244
remote sensing, xvii, xix, 86, 132
renewable energy, xviii, 4, 19, 21, 28, 37, 46, 83–84, 96, 131, 135, 143, 146, 156, 175, 186, 188, 227, 245–246, 251, 276–277, 295, 299–300, 304, 306–307, 309, 311, 313
    renewable energy sources (RES), 48–50, 127–128, 130, 171, 173, 188, 206–207, 276–279
resistance, 91, 93, 220, 299–301, 307, 309
rice, 85, 251, 261
roughages, 63

S

salinity, 89, 228
scale, xvii–xviii, 6, 9, 12, 15, 17, 26–27, 47, 51–55, 57, 59–63, 65–67, 69, 71–73, 75–79, 81, 84–85, 102–104, 106–107, 120–121, 127, 132, 135–136, 143, 147, 171, 216, 238, 252, 276, 278, 282–283, 293, 296, 302, 304, 309
seasonal, xviii–xix, 35, 40, 93, 103, 111, 129–133, 135–137, 139, 141–147, 202, 274

seasonal amplitude, 143
seasonal capacity, 135
seasonal energy storage, vii, ix, xix, 131–133, 135–137, 139, 141, 143, 145–147
seasonal storage requirement, 136, 143
seasonal variability, 143
semiconductor, 206, 224
senescence, 103, 121, 125
sensitivity, 16–17, 20, 65, 67, 75–76, 254, 270–271
    sensitivity analysis, 16–17, 20, 65, 75–76, 270–271
sewer system, 154, 156, 170
sludge, 155, 173, 200, 205–206, 209, 213, 235, 238, 240
socio-economic, 300
soil
    soil carbon, xvi, 11, 48, 50, 53, 82, 126, 128, 130, 259, 272, 275, 277
    soil depth, 45, 88–89, 91
    soil fertility, 103–104, 110, 117, 126–127, 209, 283, 292, 298
    soil organic carbon (SOC), 21, 34–36, 39, 45–47, 50, 173, 258–259, 261, 267, 273, 284, 295
    SOC sequestration, 45–46
    soil quality, 130, 132, 256
    soil texture, 88–89, 91, 94
solar radiation, 189–192, 194–195, 199, 225–227
soybean, 23, 45, 51, 58, 229, 263
spatial variance, 67, 69
stakeholder, 282, 284, 286, 288–290, 292, 294, 297–304, 309, 314
standard deviation, 67, 115, 117
stripping, 152
sugar beet, 18
sugarcane, xx, 48, 249–252, 255–257, 261, 263, 265–267, 269–279
sulfuric acid (H_2SO_4), 161–163, 167

sunflower, 18–19, 21, 23
sustainability assessment, 289, 307
switchgrass, xvi, xviii, xx, 34–37, 39–
46, 48–49, 101, 103, 120–121, 123,
125–126, 128–130, 135, 145, 147,
251–252, 256–260, 263, 265–267,
269–273, 276–278

T

temperature, xix–xx, 88–91, 94, 162,
164, 168–169, 176–185, 201–202,
209, 211–212, 216, 220, 222–223,
226–228, 232, 234–235
terrain, 69, 86, 88–89, 91, 95, 106
Thermomyces lanuginosus, 169
tillage, xvi, 34–35, 45–46, 50, 53, 76,
80
toxicity, 27
triglycerides, 159–160

U

uncertainty, xvi, 4, 24, 26, 29, 54, 132,
134, 144, 259, 271, 275, 296, 304

V

volatile organic compounds (VOC),
159, 213

W

waste management, 6, 292
waste treatment, 211
water, xix, 11, 89, 128, 132, 144, 153–
155, 160–161, 163–164, 167–169,
171–174, 176–179, 181–182, 184,
188–189, 192, 194–196, 206, 210,
212, 215–216, 218–221, 224–226,
229, 232, 235, 240–241, 243, 276,
285–286, 293, 295–298, 301, 306, 308
wastewater, 37, 172–173, 229, 235,
240
water intensity, 297
water quality, 11, 285, 297
water resources, 132, 144, 297
water solubility, 179
water storage capacity, 89
wheat, 18, 51, 58, 75, 78, 251, 282,
292, 295, 297

X

xylan, xix–xx, 175–177, 179–186
xylan removal, 176
xylose, 37, 107, 176, 210

Y

yeast, 255–256, 260, 271, 276
yield, xvi, xviii–xix, 10, 12, 15–16,
25–27, 34–36, 38, 40–43, 45–47,
49, 58, 62–63, 91, 96, 101–104,
106–108, 110–130, 133, 135, 144,
146, 152, 163, 168–169, 171, 177,
179–181, 184, 210, 229, 252,
255–258, 260, 271, 273, 285, 294,
296–297, 300–301, 309

Z

zeolite, 163, 220, 222–223